Competency Based Mathematics

for

Secondary Schools

Book 5

(MODULES 18 TO 20)

Nji Emmanuel Ndi
GBHS Mankon - Bamenda
North West Region Cameroon
Tel: (+237) 676 684 050
Email: manuelndike@gmail.com

First Edition

Printed by CreateSpace, an Amazon.com Company

EStore address: www.CreateSpace.com/7056061

Available from Amazon.com, CreateSpace.com, and other retail outlets

Available on Kindle and other retail outlets

Books by Nji Emmanuel Ndi

Complete Ordinary Level Mathematics Passport

Rudiments of Ordinary Level Mathematics (Second Edition)

Advanced Level Mathematics Key Facts

Competency Based Mathematics for Secondary Schools Book 1

Competency Based Mathematics for Secondary Schools Book 2

Competency Based Mathematics for Secondary Schools Book 3

Competency Based Mathematics for Secondary Schools Book 4

Competency Based Mathematics for Secondary Schools Book 5

DEDICATION

Dedicated to all emerging and
emergent Societies

Table of Contents

Acknowledgement

My deepest gratitude goes to God Almighty for the inspiration and for the strength.

Many thanks go to Mme. Mbuameh Daisy and Mr. Mburubah Walters for their critical proof reading of the typescript and for offering very useful suggestions which went a long way to reshape the work, the North West Regional Pedagogic Inspector for Mathematics Mr. Nfor Samuel Ndi who preview the initial manuscript and gave ample advice, which went a long way to reshape the document. I heartily thank the Former North West Regional Pedagogic Inspector for Mathematics Mr. Nji Samuel Tatah who made a very commendable effort to edit the Mathematics content of the book. I cannot forget the last minute encouragements and advice which the National inspector of Mathematics Mme Babila Emilia inspired me with. I equally pay much tribute to my students on which this material was tested. I cannot end here without thanking my sweet heart Nji Irene Nfih and my Children who encouraged and supported me in one way or the other during the course of the work.

Many thanks go to the WAEC and the CGCE Board for allowing their past questions to be used directly or indirectly.

Nji Emmanuel Ndi
G.B.H.S. Mankon, Bamenda
North West Region
Cameroon
TEL: (+237)76684050
E-mail: manuelndike@gmail.com

How to Use this Book

This book is a series written in a very special way with different sections boxed and represented by special symbols as follows.

? **Brainstorming Exercise**

Example

Real life Examples

Exercise

Skill Building Exercise

Discussion Exercise

Integration Activity

Investigative Activity

Multiple Choice Exercise

Review Exercise

Group Activity

The various sections represented by different symbols are out to facilitate navigation through the book. By investing enough time and energy in each section both students and teachers will realize that their speed and understanding will be greatly enhanced.

The brain storming exercises are aimed at provoking and invoking the learners' minds to prepare them for the task at hand. The teacher is highly encouraged to orally question the students during lessons using questions under this section.

The investigative exercises are meant to give the learner ample opportunity to experiment and self discover facts and concepts and develop methods and skills without being told.

The group activities and discussion exercises are aimed at developing a team spirit in the learners.

Many well designed examples are vividly used and solved to facilitate the learner's understanding by showing the necessary steps required for a particular solution. There are a good number of real life examples which point out the application of the subject matter in real life situations. The student is advised to study these examples very carefully.

There are many well graded exercises and skill building exercises to test the level of understanding of the learner and to facilitate skill development in the learner. The student is advised to attempt all the questions as each question may have its own technique.

Many integration activities have been designed to unify groups of sub topics, topics or modules in some cases.

Where necessary review exercises have been given to help the learner retain the skills acquired in the earlier sections.

Finally each topic ends with a good number of multiple choice questions. In each question only one of the alternatives is correct. Write down the letter corresponding to the correct answer.

For greatest achievement, the learner is advised to study regularly what he does not know and work without fear of making mistakes whether with the teacher or during group work.

By consistently and systematically going through this course as instructed, the learner will be overwhelmed with the competencies acquired at each level and at the end of the course.

Notation Used in this Book

$\{...\}$	The set of elements…or the unordered list with elements…
$n(A)$	The number of elements in set A
$\{x: \quad \}$	The set of all x such that
$\in$	Is an element of …
$\notin$	Is not an element of …
$\{\ \}$ *or* $\emptyset$	The empty set.
$\mathscr{E}$	The universal set.
$\cup$	The union of…
$\cap$	The intersection of…
$\subseteq$	Is a subset of …
$\subset$	Is a proper subset of …
$A \backslash B$	The difference between the sets A and B.
$(a, b, c, ...)$	An ordered list of elements a, b, c, …
$\{a, b, c, ...\}$	The set or an unordered list of elements a, b, c, …
$\mathbb{Z}$	The set of integers, $\{0, \pm 1, \pm 2, \pm 3, \pm 4, ...\}$
$\mathbb{N}$	The set of all positive integers and zero, $\{0,1,2,3,4, ...\}$
$\mathbb{Z}^+$	The set of positive integers $\{+1, +2, +3, +4 ...\}$
$\mathbb{Q}$	The set of rational numbers
$\mathbb{Q}^+$	The set of positive rational numbers
$\mathbb{R}$	The set of all real numbers $\{x: x \in \mathbb{R}\}$
$\mathbb{R}^+$	The set of all positive real numbers $\{x \in \mathbb{R}: x > 0\}$
$f(x)$	f *of* x or the image of x under the function f
f^{-1}	The inverse function of the function f
fg or $f \circ g$	The function f of the function f
$=$	Is equal to
$\neq$	Is not equal to
$\approx$	Is approximately equal to
$<$	Is less than
$>$	Is greater than
$\not<$	Is not less than
$\not>$	Is not greater than

$\leq$	Is less than or equal to
$\geq$	Is greater than or equal to
$a < x < b$ or $]a,b[$ or (a, b)	An open interval on the number line
$a \leq x \leq b$ or $[a, b]$	A closed interval on the number line
$\{x : a < x < b\}$	The set of elements x such that a is less than x and x is less than b
a	The vector **a**
AB	The vector represented in magnitude and direction by AB
$\|x\|$	The modulus or absolute value of x i.e. $\{x$ for $x > 0, -x$ for $x < 0, x \in \mathbb{R}\}$
$\boldsymbol{a} \cdot \boldsymbol{b}$	The dot or scalar product of the vectors **a** and **b**
A^{-1}	The inverse of the non-singular matrix A
A^T	The transpose of the matrix A
$\lg x$ or $\log x$	The common logarithm of x
x^n	The number x, raised to the power n
$\propto$	Is proportional
∞	Infinity
$\sqrt{}$	The positive square root
$-\sqrt{}$	The negative square root
$\sqrt[n]{a}$	The nth root of a
$p:$	The statement or preposition p
T or 1 in truth tables	True
F or 0 in truth tables	False
$\sim p$ or p' or $\neg p$	The negation of a statement p
$p \wedge q$	The conjunction of the statements p and q
$p \wedge q$	The disjunction of the statements p and q
$A \cap B$ or $\{x : x \in A \wedge x \in B\}$	The intersection of sets A and B.
$A \cup B$ or $\{x: x \in A \vee x \in B\}$	The union of sets A and B.
$p \Rightarrow q$ or $p \rightarrow q$	p implies q or p is sufficient for q or p only if q or q is necessary for p

$p \Leftrightarrow q$ or $p \leftrightarrow q$ or p iff q	p is a necessary and sufficient condition for q or p implies and is implied by q or p if and only if q
$\forall x$	For all or for every element x
$\exists x$	There exists or for at least one or for some element x
$\exists ! x$	There exists one and only one element x
$\equiv$	Is equivalent or is congruent to
$///$	Is similar to
$\perp$	Is perpendicular to
$\parallel$	Is parallel to
$G = (V, E)$	The graph of the set V of vertices together with the set E of edges
$D = (V, A)$	The directed graph (digraph) of the set V of vertices and the set A of ordered edges
$G = (V, E, A)$	The mixed graph of the set of vertices V, unordered edges E and ordered edges A.
$n :=$	The store n takes the value…
$^\circ$	Degree
$^\circ C$	Degrees Celsius
$^\circ F$	Degrees Fahrenheit

Module 18

Plane Geometry

Family of Situations

Module 18 is an extension of module 2, 6, 11, 16 and 18. At the end of the module; the student is expected to have acquired many more competencies within the **families of situations** *'Representation and transformation of Plane Shapes within the Environment'.*

Categories of Action

The categories of action for module 18 include:
1. Perception of the physical environment,
2. Production of plane shapes,
3. Transformation of the physical environment,
4. Determination of measures and position within the physical environment.

Credit

The module is expected to be covered within 11 weeks teaching 4 periods of 50 minutes per week (or within 44 periods).

Topic 1

COORDINATE GEOMETRY

Objectives

At the end of this topic, the learner should be able to:

1. Divide a line segment internally or externally in a given ratio.
2. Find the distance between two points.
3. Find the equation of a straight line with given conditions.
4. Justify that two lines are parallel or perpendicular.
5. Solve simultaneous equations graphically.
6. Make a table of values for a quadratic function and draw the graph.
7. Find the gradient of a tangent to a curve at a point.
8. Find the coordinates of particular points. (maximum, minimum, point of inflexion, point of intersection with the axes etc)

1.1   Point Dividing a Line Segment

Consider the line segment AB in the figure below. P divides AB internally because P is between A and B while Q divides AB externally because Q is not between A and B.

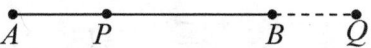

1.2   The Midpoint of a Line

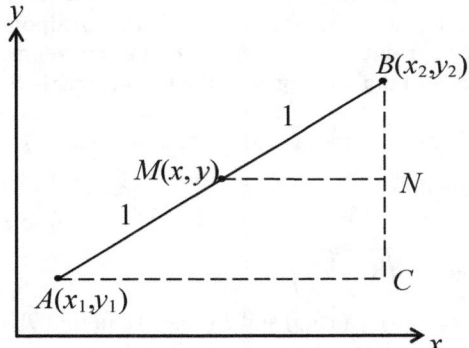

? Brainstorming Exercise

In the figure above, $M(x, y)$ is the midpoint between the points $A(x_1,y_1)$ and $B(x_2,y_2)$. MN and AC are parallel to the x-axis and BC is parallel to the y-axis.
1.   What can you say about triangles ABC and MBN?
2.   In what ratio does M divide AB?
3.   Write down BN, BC, MN and AC in terms of x, x_1, x_2, y, y_1 and y_2.
4.   Write down two relationships between the sides of triangles ABC and MBN.

Clearly, triangles ABC and MBN are similar and using similar triangles

$$\frac{y_2 - y}{y_2 - y_1} = \frac{1}{2} \Leftrightarrow y = \frac{y_2 + y_1}{2} \quad \text{and} \quad \frac{x_2 - x}{x_2 - x_1} = \frac{1}{2} \Leftrightarrow x = \frac{x_2 + x_1}{2}.$$

Therefore, if $A(x_1,y_1)$ and $B(x_2,y_2)$ are any two points, the midpoint $M(x, y)$ of the line segment is given by

$$M(x, y) = \left(\frac{x_1 + x_2}{2}, \frac{y_1 + y_2}{2} \right)$$

3

Example

Find the mid-point between $A(8,2)$ and $B(-2,-6)$.

Solution

$$M(x,y) = \left(\frac{x_1 + x_2}{2}, \frac{y_1 + y_2}{2}\right) = \left(\frac{8 + (-2)}{2}, \frac{2 + (-6)}{2}\right) = (3,-2)$$

Exercise 1:1

1. In the following, find the coordinates of the midpoint of the line segment AB
   (a) $A(-2, 7)$ and $B(-2, -11)$ (b) $A(-5, -4)$ and $B(-10, -4)$
2. Write down formulae, which we can use to find the midpoint of
   (a) A vertical line segment   (b) A horizontal line segment
3. Find the coordinates of the midpoints of the line segments between the following points
   (a) $(6, 0)$ and $(10, 2)$ (b) $(-5, 6)$ and $(6, -5)$

   (c) $(0, -7)$ and $(-7, 0)$ (d) $\left(\frac{3}{4}, -\frac{5}{3}\right)$ and $\left(\frac{3}{4}, \frac{2}{3}\right)$

   (e) $\left(\sqrt{2}, -\sqrt{3}\right)$ and $\left(\sqrt{8}, \sqrt{75}\right)$
4. Given the points, $P(7,13)$, $Q(10,5)$ and $R(-6,-3)$ and that T is the midpoint of QR, calculate the distance PT.

1.3   Internal Division

Suppose $P(x, y)$ divides AB internally in the ratio $m{:}n$, where A is the point (x_1,y_1) and B is the point (x_2,y_2) as shown in the figure below.
    Then using similar triangles

$$\frac{y_2 - y}{y_2 - y_1} = \frac{n}{m+n} \Leftrightarrow y = \frac{my_2 + ny_1}{m+n}$$

Similarly;

$$\frac{x_2 - x}{x_2 - x_1} = \frac{n}{m+n} \Leftrightarrow x = \frac{mx_2 + nx_1}{m+n}$$

Therefore,

$$P(x, y) = \left(\frac{mx_2 + nx_1}{m+n}, \frac{my_2 + ny_1}{m+n} \right)$$

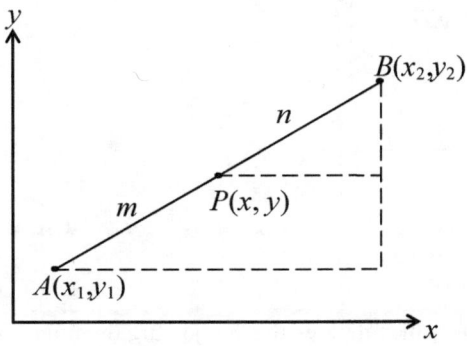

1.4   External Division

Suppose $Q(x, y)$ divides $A(x_1,y_1)$ and $B(x_2,y_2)$ externally in the ratio $m:n$, as shown in the figure below.

Again using similar triangles,

$$\frac{y - y_2}{y - y_1} = \frac{n}{m} \Leftrightarrow y = \frac{my_2 - ny_1}{m-n}$$

Similarly,

$$\frac{x - x_2}{x - x_1} = \frac{n}{m} \Leftrightarrow y = \frac{mx_2 - nx_1}{m-n}$$

Therefore,

$$P(x, y) = \left(\frac{mx_2 - nx_1}{m-n}, \frac{my_2 - ny_1}{m-n} \right)$$

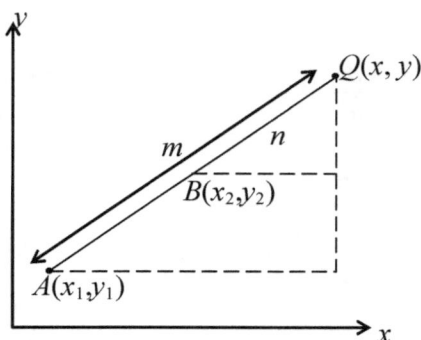

 Example

Find the coordinates of the point which divides the line joining the points $A(-2,5)$ and $B(4,2)$ in the ratio $2:1$. (a) internally   (b) externally.

Solution

Let the coordinate of the point that divides AB be (x,y). Then,

(a) Internally $(x,y) = \left(\dfrac{mx_2 + nx_1}{m+n}, \dfrac{my_2 + ny_1}{m+n} \right) = \left(\dfrac{2(4)+1(-2)}{2+1}, \dfrac{2(2)+1(5)}{2+1} \right)$

$\Rightarrow (x,y) = (2,3)$

(b) Externally $(x,y) = \left(\dfrac{mx_2 - nx_1}{m-n}, \dfrac{my_2 - ny_1}{m-n} \right) = \left(\dfrac{2(4)-1(-2)}{2-1}, \dfrac{2(2)-1(5)}{2-1} \right)$

$\Rightarrow (x,y) = (10,-1)$

A closer look at the formulae reveals that,

(a) For internal division, we can use formula ① directly.

(b) For external division, we can write the ratio as $m: -n$ and use it in formula ① directly.

 Example

The point $P(x,y)$ divides the line AB in the ratio $3:2$ where A is the point $(-1, 6)$ and B is the point $(3, -2)$. (a) Internally   (b) externally.
Find the coordinates of the point P in each case.

Solution

$$\text{Internally } (x, y) = \left(\frac{mx_2 + nx_1}{m+n}, \frac{my_2 + ny_1}{m+n} \right)$$

(a) For internal division, the ratio is $m : n = 3 : 2$

$$\Rightarrow P(x, y) = \left(\frac{3(3) + 2(-1)}{3+2}, \frac{3(-2) + 2(6)}{3+2} \right) = \left(\frac{7}{5}, \frac{6}{5} \right)$$

(b) For external division, the ratio is $m : n = 3 : -2$

$$\Rightarrow P(x, y) = \left(\frac{3(3) + (-2)(-1)}{3+(-2)}, \frac{3(-2) + (-2)(6)}{3+(-2)} \right) = (11, -18)$$

 ## Exercise 1:2

Find the coordinates of the point, which divides AB in the given ratio in each of the following cases.

1. $A(2, 4), B(-3, 9); 1:4$ internally
2. $A(-3, -4), B(3, 5); 3:1$ externally
3. $A(1, 5), B(8, -2); 4:3$
4. $A(5, 2), B(-2, 8); 3:-2$

 ## Investigative Activity

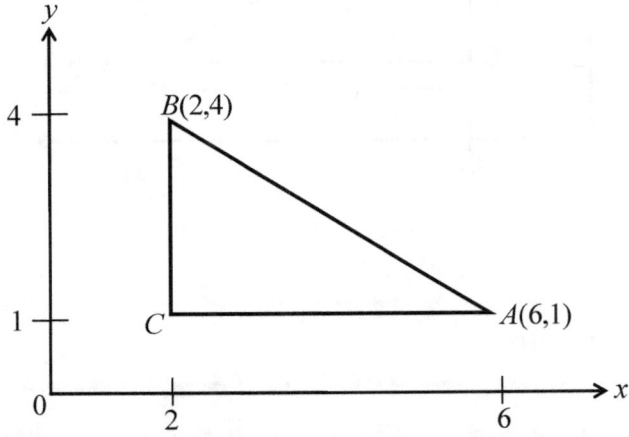

In the figure above, $A(6,1)$ and $B(2,4)$ are two points, AC are parallel to the x-axis and BC is parallel to the y-axis.

1. Write down the coordinates of the point C.
2. Find the distances AC and BC.
3. Applying the Pythagoras theorem, use the distances AC and BC to find the length of AB.
4. Evaluate the ratio $\frac{BC}{AC}$.
5. Given $A(x_1, y_1)$ and $B(x_2, y_2)$ questions 1 to 4 above.

1.5   The Distance between Two Points

From the above investigative activity, we see that if $A(x_1, y_1)$ and $B(x_2, y_2)$ are any two points as shown in the diagram below, we can find the distance AB between A and B using the Pythagoras theorem. Thus,

$$AB = \sqrt{(x_2 - x_1)^2 + (y_2 - y_1)^2}$$

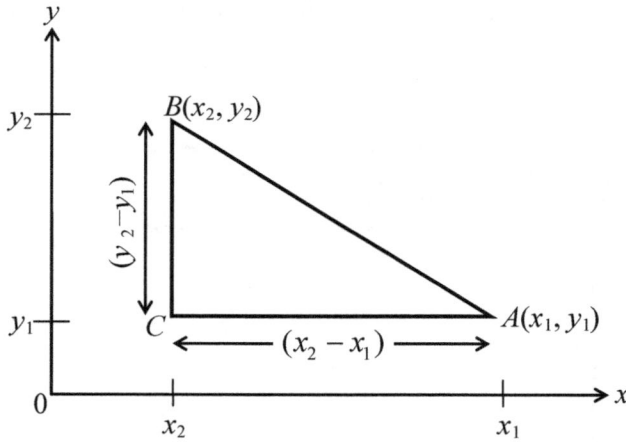

 Example

1. Plot on a graph the points $A(1,0)$ and $B(5,3)$. By completing the right-angled triangle with AB as one of the sides, calculate the length of the line AB.

Solution

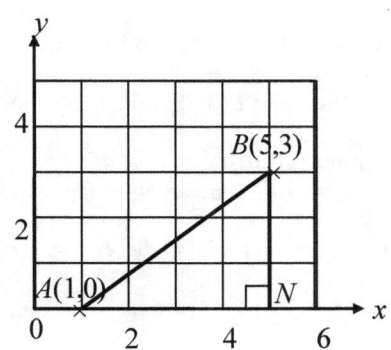

$$AB^2 = AN^2 + BN^2 = (5-1)^2 + (3-0)^2 = 25$$
$$\Rightarrow AB = \sqrt{25} = 5 \text{ units}$$

2. On a Cartesian plane, plot the points $Q(8,6)$ and $R(16,12)$ and calculate the distance between them.

Solution

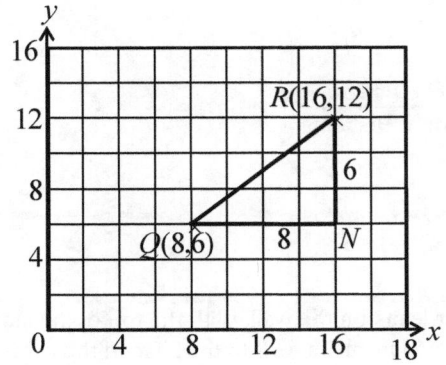

$$QR^2 = QN^2 + RN^2 = (16-8)^2 + (12-6)^2 = 8^2 + 6^2 = 100$$
$$QR = \sqrt{100} = 10 \text{ units}$$

 ## Exercise 1:3

1. Find The distance between the following pairs of points;
    (a) $(-2, 3)$ and $(-1, 4)$ (b) $(1, 2)$ and $(6, 14)$ (c) $(3, -4)$ and $(0, 0)$
    (d) $(4, -2)$ and $(-10, 1)$ (e) $(-12, -15)$ and $(-14, -15)$

2. Triangle ABC has vertices $A(-5,5)$, $B(7,0)$ and $C(12,5)$. Calculate the length of each side.

3. Given that the triangle with vertices $(0,6)$, $(k,-k)$ and $(-6,0)$ is equilateral. Find the value of k.

4. The point $A(x, y)$ is equidistant from the point $B(3,3)$ and $C(7,5)$. Find an equation, which connects x and y.

5. Calculate the distance between the points $A(at, a)$ and $(-at, at^2)$.

6. The point (x, y) is 4 units from the point $(3,4)$. Find an equation, which connects x and y.

7. The points $A(1,2)$, $B(5,-6)$ and $C(k, k)$ are such that angle ABC is a right angle. Find the value of k.

8. Given that the triangle with vertices $(5,7)$, $(14, -5)$ and $(x, -5)$ is equilateral. Find the possible values of x.

1.6   Gradient

The **gradient** m of a slope is the measure of the steepness of a slope. The gradient of a slope compares the vertical distance (rise) and the horizontal distance (run) when one ascends a slope.

$$m = \frac{\text{Rise}}{\text{Run}} \quad \text{or} \quad m = \frac{\text{Vertical Distance}}{\text{Horizontal Distance}} \quad \text{or} \quad m = \frac{v}{h}$$

Example

A wall is 6 m tall and a ladder leans on the wall, with the top of the ladder touching the top of the wall. If the distance from the wall to the base of the ladder is 2 m, calculate the gradient of the ladder.

Solution

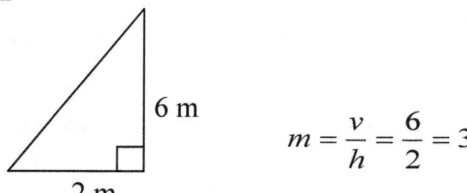

$$m = \frac{v}{h} = \frac{6}{2} = 3$$

6 m

2 m

1.7   The Gradient of a Straight line

From the above investigative activity we see that if $A(x_1, y_1)$ and $B(x_2, y_2)$ are any two points as shown in the diagram below, then from the definition of gradient, the gradient of the straight line passing through the points A and B is,

$$m = \frac{v}{h} \Rightarrow m = \frac{y_2 - y_1}{x_2 - x_1}$$

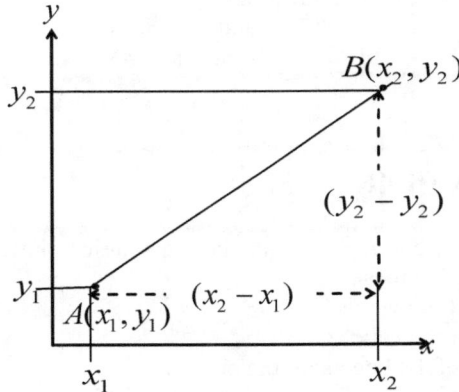

 Example

1. Find the gradient of the line joining the points $P(4,11)$ and $Q(7,2)$.

    Solution
    $$m = \frac{y_2 - y_1}{x_2 - x_1} = \frac{2 - 11}{7 - 4} = \frac{-9}{3} = -3$$

2. Calculate the steepness of the line joining the points $A(2,1)$ and $B(-5,3)$.

    Solution
    $$m = \frac{y_2 - y_1}{x_2 - x_1} = \frac{3 - 1}{-5 - 2} = -\frac{2}{7}$$

Exercise 1:4

1. Find the gradient of the line, which passes through each of the following pair of points
   (a) (−4, 2) and (3, 5)   (b) (−4, 7) and (−4, 6)   (c) (3, 3) and (−3, −3)
   (d) (−3, −3) and (5, 3)   (e) $\left(\dfrac{1}{2}, \dfrac{1}{3}\right)$ and $\left(\dfrac{1}{4}, \dfrac{1}{5}\right)$ (f) (0, 0) and (−4, −3)

2. Given that the gradient of the line which passes through the points (7,−6) and (k, 2) is 2. Find the value of k?

3. Given that the gradient of the line, which passes through the points (k, 5), and (0,−11) is k. Find the value of k.

4. Calculate the gradient of the line joining the points
   (a) $A(1,3)$ and $B(−3,9)$ (b) $P(7,4)$ and $Q(−5,−6)$
   (c) $R(3,5)$ and $S(6,9)$ (d) $X(−3,8)$ and $Y(6,4)$

 Investigative Activity

1.   Plot the points $P(1,3)$, $Q(3,7)$ and $R(−2,−3)$ and connect these points.
     (a) Make a remark concerning these points.
     (b) Where does the line cut the y-axis?
     (c) Where does the line cut the x-axis?
     (d) Calculate the gradient of the line using the points.
          (i) P and Q (ii) P and R (iii) Q and R
     (e) What conclusion do you draw and what does this suggest?
2.   Given that the gradient of a straight line which passes through the points $P(x, y)$ and (o, c) is m. Write an equation involving x, y, o, c and m.
3.   A straight line passes through the points $A(x_1, y_1), B(x_2, y_2)$ and $P(x, y)$.
     (i) Find the gradient m of the line using the points
          (a) $A(x_1, y_1)$ and $B(x_2, y_2)$ (b) $A(x_1, y_1)$ and $P(x, y)$.
     (ii) Write an equation involving x_1, y_1, x_2, y_2 and m.
     (iii) Write an equation involving x_1, y_1, x, y and m.
     (iv) Write an equation involving x_1, y_1, x_2, y_2 and x, y.

1.8   Intercepts

We call the y-coordinate of the point where a line cuts the y-axis the y-**intercept** or the **intercept with the y-axis** and we call the x-coordinate of the point where a line cuts the x-axis the x-**intercept** or the **intercept with the x-axis**. Where a line cuts the y-axis, $x = 0$ and where a line cuts the x-axis, $y = 0$.

From the investigative exercise above, we see that we can calculate the gradient of a straight line using the coordinates of any, two points that lie on the line.

1.9 The Equation of a Straight Line

The various forms of the equation of a straight line are; the **gradient and one point form**, the **gradient–intercept form**, the **double intercept form**, the **two-point form** and the **general form**.

1.10 Gradient and one point form

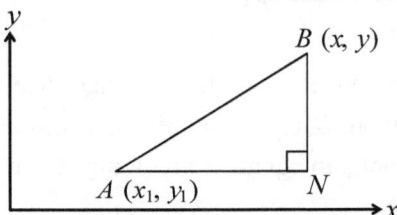

From the investigative exercise above, given that a straight line has gradient m and passes through the points A (x_1, y_1) and B (x, y) as shown in the figure above, then, an equation involving x_1, y_1, x, y and m is

$$m = \frac{y - y_1}{x - x_1} \quad \dots\dots\dots\dots\dots \text{①}$$

Rearranging equation ①,

$$y - y_1 = m(x - x_1) \quad \dots\dots\dots\dots\dots \text{②}$$

This is the equation of a Straight Line known as the **Gradient and one point form**. This form of the equation of a straight line is useful when, the gradient of the line and one point on the line are known.

 Example

Find the equation of the line, which passes through the point $(2, -5)$ and whose gradient is -2.

Solution

$$y - y_1 = m(x - x_1)$$

$x_1 = 2, y_1 = -5$ and $m = -2$

$$y - (-5) = m(x - 2) \qquad \Rightarrow \quad y = -2x - 1 \text{ is the equation of the line.}$$

13

1.11 Gradient – Intercept form

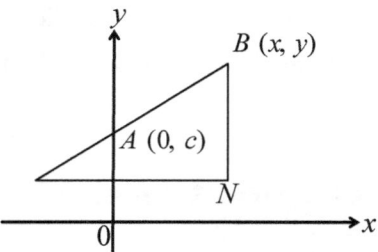

From the investigative exercise above, given that a straight line which has gradient m passes through the point $B(x, y)$ and cuts the y-axis at the point $A(0, c)$ as shown in the figure above, then, an equation involving c, x, y and m is

$$\frac{y-c}{x-0} = m$$

We can rearrange this as

$$y = mx + c \dots\dots\dots\dots\dots③$$

Equation ③ is known as the **Gradient-intercept form** of the equation of a Straight Line and is useful when we know the gradient of the line and the intercept with the y-axis.

For this form, the gradient is m and we can obtain the intercepts by substituting $x = 0$ or $y = 0$ in the equation.

Thus, when $x = 0$, $y = c$ and when $y = 0$, $x = -\dfrac{c}{m}$.

 Example

A line whose gradient is $-\dfrac{3}{4}$, is known to cut the y-axis at the point $(0, 5)$. Find the equation of the line.

Solution

$$y = mx + c, \quad m = -\frac{3}{4} \text{ and } c = 5 \Rightarrow y = -\frac{3}{4}x + 5$$

1.12 The Double Intercept Form

From the investigative exercise above, given that the straight line which cuts the x- and y-axes at the points $(a,0)$ and $(0,b)$ respectively and passes through any other point $B(x, y)$ as shown in the figure below, then, an equation involving a, b, x and y is

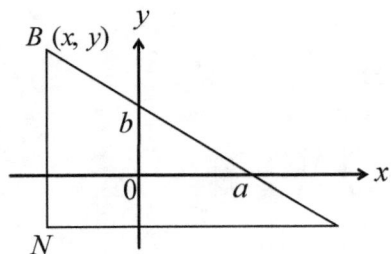

$$\frac{y-0}{x-a} = \frac{b-0}{x-a}$$

Cross multiplying and rearranging we have,

$$\frac{x}{a} + \frac{y}{b} = 1 \quad \dots\dots\dots\dots\dots\dots\dots\text{④}$$

This form of the equation of a straight line which makes use of the x and y intercepts is known as the double intercept form and is useful when the intercepts with the x- and y-axes are known.

1.13 Two point form

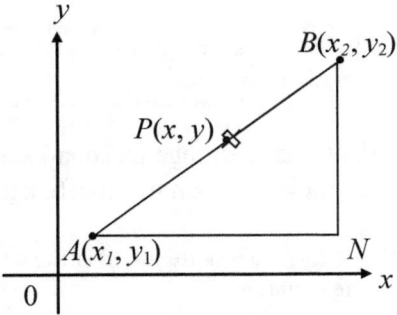

From the investigative exercise above, given that a straight line passes through the points $A(x_1, y_1)$ and $B(x_2, y_2)$ and $P(x, y)$ as shown in the figure below, then, the equation involving x_1, y_1, x_2, y_2, x and y is

$$\frac{y - y_1}{x - x_1} = \frac{y_2 - y_1}{x_2 - x_1} \quad \dots\dots\dots\dots\dots\dots(5)$$

This form of the equation of a straight line is known as the two-point form and is useful when two points that lie on the straight line are known.

1.14 General form

Cross-multiplying equation ⑤ becomes,

$$(y - y_1)(x_2 - x_1) = (y_2 - y_1)(x - x_1)$$

$$yx_2 - yx_1 - y_1x_2 + y_1x_1 = y_2x - y_2x_1 - y_1x + y_1x_1$$
$$(y_1 - y_2)x + (x_2 - x_1)y + (x_1y_2 - x_2y_1) = 0$$

Substituting a, b and k for the constants $x_2 - x_1$, $y_1 - y_2$ and $x_1y_2 - x_2y_1$ leads to

$$ax + by + k = 0 \quad \dots\dots\dots\dots\dots\dots(6)$$

Equation ⑥ is known as the general form of the equation of a straight line.

1.15 Obtaining the gradient and the intercepts from an equation of a straight line

> ? **Brainstorming Exercise**
>
> Explain how you will obtain the gradient of a straight line and the intercept with each axis from each form of the equations of a straight line.

1. To find the gradient of a straight line rearrange its equation so that the coefficient of y is unity. The gradient is then the coefficient of x in the rearranged equation.
2. To find the intercept with the x-axis substitute $y = 0$ in the equations of the straight line and solve for x in the equation.
3. To find the intercept with the y-axis substitute $x = 0$ in the equations of the straight line and solve for y in the equation.

Example

Given the line $\dfrac{x}{2} + \dfrac{y}{5} = 1$, find

(i) the intercept with the x-axis   (ii) the intercept with the y-axis

(iii) The gradient of the line.

Solution

(i) Intercept with the x-axis is 2, when $y = 0$.

(ii) Intercept with the y-axis is 5, when $x = 0$.

(iii) $m = -\dfrac{5}{2}$

The following table is a summary of the four forms of the equation of a straight line.

Form	Gradient	Intercept	
		x-axis	y-axis
Gradient/ Intercept Form $y = mx + c$	m	c	c
General Form $ax + bx + k = 0$	$-\dfrac{a}{b}$	$-\dfrac{k}{a}$	$-\dfrac{k}{b}$
Double/Intercept Form $\dfrac{x}{a} + \dfrac{y}{b} = 1$	$-\dfrac{b}{a}$	a	b
Gradient/ one point form $y - y_1 = m(x - x_1)$	m	$\dfrac{mx_1 - y_1}{m}$	$y_1 - mx_1$

1.16 Parallel and Perpendicular Lines

Investigative Activity

On square or graph paper, draw the straight lines with the given equations.

$L_1: y = 2x + 5$, $L_2: y - 2x - 6 = 0$, $L_3: 2y + x - 6 = 0$

1. From your graph, which of the lines are:
    (a) Parallel?   (b) Perpendicular?

2. State the gradients m_1, m_2 and m_3 of the lines.
3. Deduce a relationship between the gradients of parallel lines.
4. Evaluate (a) m_3m_1 (b) m_3m_2.
5. Deduce a relationship between the gradients of Perpendicular lines.

From the investigative activity above, we can see that if the lines l_1 and l_2 have gradients m_1 and m_2 respectively, then:
(a) The lines are parallel if and only if their gradients are equal.

$$l_1 \parallel l_2 \Leftrightarrow m_1 = m_2$$

(b) The lines are perpendicular if and only if the product of their gradients is -1.

$$l_1 \perp l_2 \Leftrightarrow m_1m_2 = -1$$

 Example

1. Find the equation of the straight line which passes through the point $(-2, 5)$ and is perpendicular to the line $y = 2x+3$.

   Solution

   Let the gradient of the line $y = 2x + 3$ be $m_1 = 2$ and that of the perpendicular line be m_2.

   Then $m_1m_2 = -1 \Rightarrow 2m_2 = -1$

   $$m_2 = -\frac{1}{2}$$

   Therefore, the equation of the perpendicular line is

   $$y - y_1 = m_2(x - x_1)$$

   Substitute $(-2, 5)$ and $m_2 = -\frac{1}{2}$ into the equation.

   $$y - 5 = -\frac{1}{2}(x - (-2))$$

   $$\therefore y - 5 = -\frac{1}{2}x - 1 \quad \text{or} \quad y = -\frac{1}{2}x + 4$$

2. A line passes through the point $(1, 3)$ and is parallel to the line $y = 3x + 2$. Find the equation of the line.

Solution

Lines are parallel so $m_1 = m_2 = 3$.

Therefore, the equation of the parallel line is

$$y - y_1 = m_2(x - x_1)$$

Substitute $(1, 3)$ and $m_2 = 3$ into the equation.

$$y - 3 = 3(x - 1)$$
$$y - 3 = 3x - 3$$
$$\therefore\ y = 3x$$

 Exercise 1:5

1. Find the equation of the line, which passes through the following pair of points.
   (a) $(4, 1)$ and $(-2, 5)$ (b) $(-6, -7)$ and $(7, 8)$
   (c) $\left(-1, \frac{3}{4}\right)$ and $\left(-\frac{1}{2}, 4\right)$ (d) $(2, 7)$ and $(-8, 5)$
   (e) $(\sqrt{2}, \sqrt{8})$ and $(-\sqrt{8}, -\sqrt{2})$

2. Find the equation of the line, which passes through the point $(6, -1)$ and is perpendicular to the line $y - 3x - 1 = 0$.

3. Find the equation of the line, which passes through the point $(3, 1)$ and is parallel to the line $2x + 5y - 4 - 0$.

4. Find the equation of the straight line, which passes through the point $(2, -1)$ and is
   (a) perpendicular to the line $3x + y = 7$. (b) parallel to the line $3x + y = 7$.

5. A line l_1 passes through the points $(1, 8)$ and $(-2, 1)$. Find the equation of the line that passes through the point $(2, 1)$ and is
   (a) perpendicular to l_1. (b) parallel to l_1.

6. A straight line has gradient $-\dfrac{12}{5}$ and passes through the point $(2, 1)$. Find its equation.

7. Given the points $X(-2, 7)$, $Y(6, -1)$ and $Z(9, 4)$ and that P is the midpoint of the line segment $[XY]$. Calculate:
   (a) the coordinates of P (b) the numerical value of $XY : PZ$.
   (c) the equation of the line YZ.

8. A straight line whose gradient is -4 passes through the point $(2, 3)$. Find the equation of the line.

9. Given that the line which passes through the points $(b, 3)$ and $(-2, 1)$ is parallel to the line, which passes through the points $(5, b)$ and $(1, 0)$. Find the value(s) of b.

10. Given that the line which passes through the points $(3m, 0)$ and $(2n, 0)$ is parallel to the line which passes through the points $(0, n)$ and $(0, 6m)$. Find a relation between m and n.

11. If the line that passes through the points $(0,0)$ and (n, m) is perpendicular to the line that passes through the points $(0,0)$ and $(-a, b)$. Find the relation between a, b, m and n.
12. Given the straight lines A: $y - 3x - 5 = 0$, B: $x = 8 - 3y$, C: $2y = 5 - 6x$. Determine which of the lines (if any)
    (a) are perpendicular   (b) are parallel   (c) pass through the origin

1.17   Corresponding, Inconsistent and Intersecting Lines

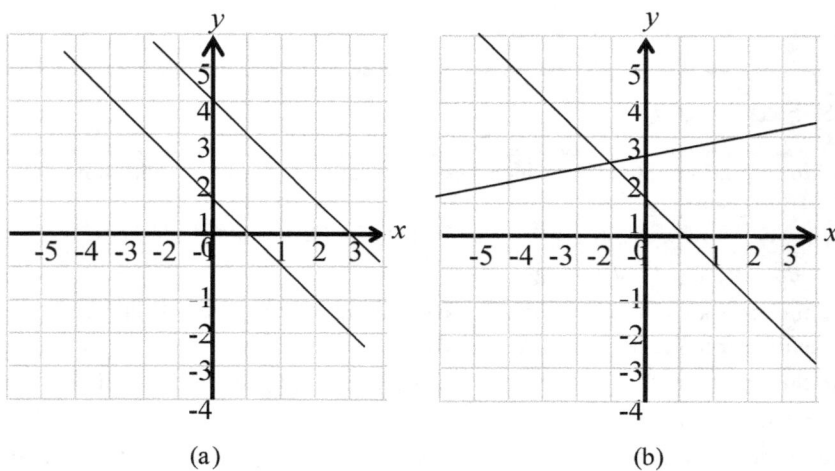

(a)   (b)

If two lines are **parallel** (Figure (a)), they do not meet anywhere. Such lines are **inconsistent**. It is also possible for two lines to be identical. If the equations $y = x + 2$ and $3y = 3x + 6$ are graphed, they will be the same. This is because one equation is simply a multiple of the other. Such lines are **corresponding lines** and their equations are **dependent**. **Intersecting lines** are lines that meet at a point as illustrated in Figure (b).

1.18 Graphical Solutions to Linear Equations

 ## Investigative Activity

1. On a square or graph paper, draw the straight lines with the following equations, making sure that each cuts the x-axis.
$$L_1: y = 2x + 5, \quad L_2: y = 2x, \quad L_3: 2y + x - 5 = 0$$
2. Where do the lines L_1, L_2 and L_3 cut the x-axis
3. State the value of x for which $y = 0$ in each of $y = 2x + 5$, $y = 2x$ and

20

$$2y + x - 5 = 0$$
4. What can you say about the point where each line cuts the x-axis and the corresponding value of x in question 2?
5. Where does the following pair of lines intersect? (a) L_1 and L_3 (b) L_2 and L_3
6. Solve the following pair of simultaneous equations.
   (a) $y = 2x + 5$ and $2y + x - 5 = 0$ (b) $y = 2x$ and $2y + x - 5 = 0$
7. Compare your answers in question 5 and 6, hence deduce a method of solving simultaneous equations using a graphical method.

From the investigative activity above we see that:
1. To solve a simple linear equation using the graphical method, plot the graph of the line. The x-coordinate of the point where the line cuts the x-axis gives the solution of the equation.
2. To solve a pair of simultaneous linear equations using the graphical method, plot the graphs of the equations on the same Cartesian axes. The coordinates of the point of intersection of the lines gives the solutions of the equations.

 Example

Solve the simultaneous equations $y = 2x$ and $y + x = 3$, using the graphical method.

Solution

$y = 2x$

x	0	1	2
y	0	2	4

$y + x = 3$

x	0	2	4
y	3	1	-1

The figure below shows the graph.
From the graph, the two lines meet at the point $(1,2)$. Therefore, the solution of the simultaneous equations is $x = 1$, $y = 2$.

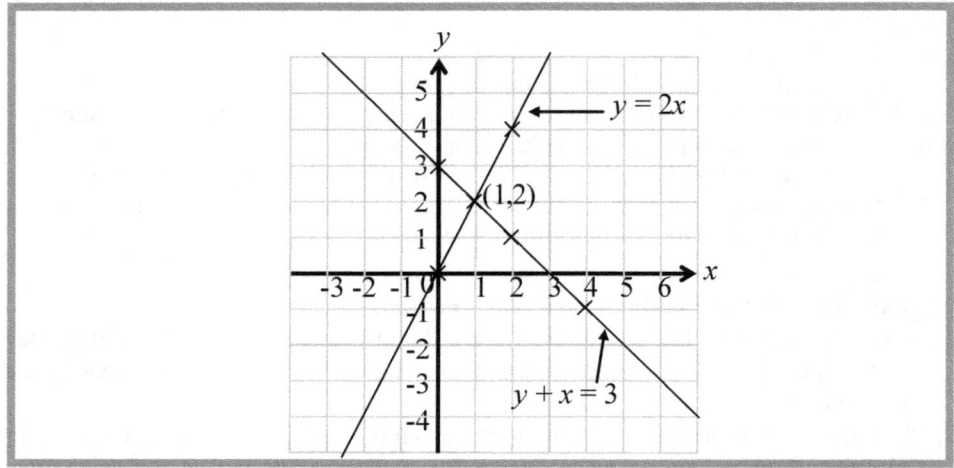

1.19 Graphs of Quadratic Functions

Recall that a quadratic function is of the form $f(x) = ax^2 + bx + c$, where a, b and c are constants and $a \neq 0$.

To plot the graph of a quadratic function, first make a table of values of y against x.

 Example

Given that $f(x) = 4 - 3x - x^2$, make up a table of values of x against $y = f(x)$, for integral values of x in the range $-5 \leq x \leq 2, x \in \mathbb{R}$. Taking 1 cm to represent 1 unit on the x-axis and 1 cm to represent 2 units on the y-axis draw the graph of $y = f(x)$.

Solution

x	$4 - 3x - x^2$	$y = f(x)$
-5	$4 + 15 - 25$	-6
-4	$4 + 12 - 16$	0
-3	$4 + 9 - 9$	4
-2	$4 + 6 - 4$	6
-1	$4 + 3 - 1$	6
0	$4 + 0 - 0$	4
1	$4 - 3 - 1$	0
2	$4 - 6 - 4$	6

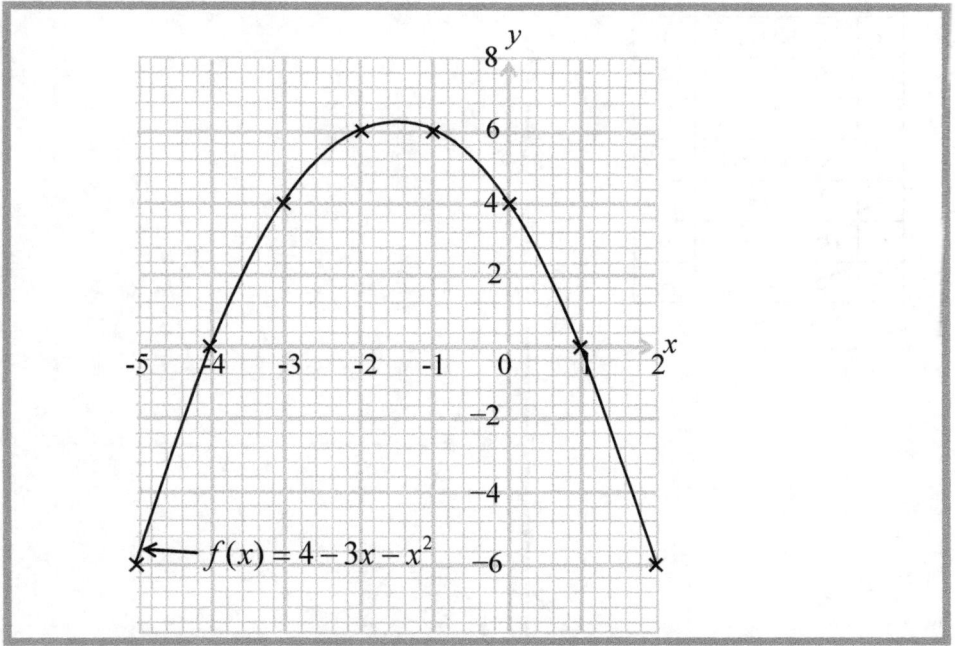

$$f(x) = 4 - 3x - x^2$$

1.20 Hints for Drawing Smooth Curves

To ensure the smoothness of a curve such as the one above,

1. Make sure that the drawing hand is always inside the curve when drawing the graph, even if it means turning the graph paper or book upside down.
2. Use a pencil to sketch the curve very lightly and retrace it when you judge that the curve is very smooth.

 ## Example

Plot the graph of $y = x^2 - 5x + 4$ for integral values of x from 0 to +5.

Solution

x	$x^2 - 5x + 4$	y
0	$0 - 0 + 4$	4
1	$1 - 5 + 4$	0
2	$4 - 10 + 4$	-2
3	$9 - 15 + 4$	-2
4	$16 - 20 + 4$	0
5	$25 - 25 + 4$	4

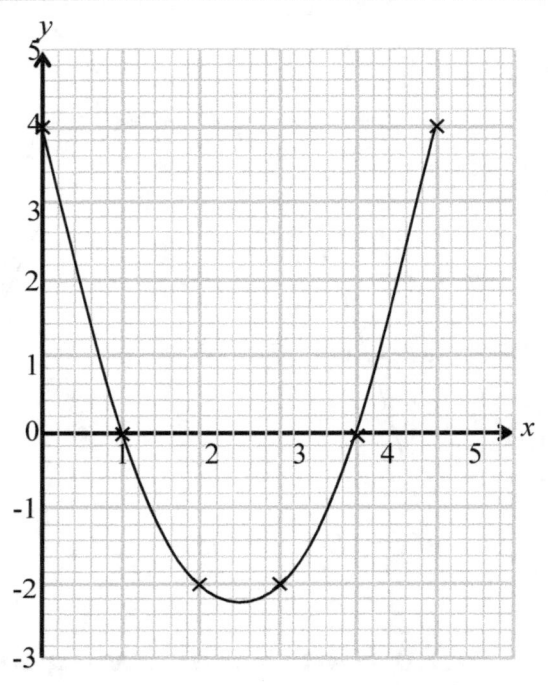

1.21 Graphical Solutions of Quadratic Equations

 Investigative Activity

1. Solve the following quadratic equations using formula or factorization method.
   (a) $4 - 3x - x^2 = 0$ (b) $x^2 - 5x + 4 = 0$
2. State the x-coordinate of the point where the graphs of $y = 4 - 3x - x^2$ and $y = x^2 - 5x + 4$ above cut the x-axis.
3. Deduce and explain a way of solving quadratic equations using the graphical method.

From the above investigative activity we see that we can find the roots of the equation $ax^2 + bx + c = 0$ by drawing the graph of $y = ax^2 + bx + c$, and looking for the x-coordinate of the point where the graph cuts the x-axis.

Exercise 1:6

1. Sketch the graph of $f(y) = 2y^2 + 2y - 15$, for integral values of y from -4 to $+4$.

2. If $f(x) = 5 + 4x - x^2$,
   (a) Construct a table of values of the function $y = f(x)$ for $-2 \le x \le 5$.
   (b) Using a scale of 2 cm for 1 unit on the x-axis and 2 cm for 2 units on the y-axis, draw the graph of $y = f(x)$ for $-2 \le x \le 5$.
   (c) Use your graph to find the range of values of x for which $5 + 4x - x^2 > 0$.

3. Solve the following equations graphically
   (i) $x^2 - 1 = 0$ (ii) $4x^2 - 9 = 0$ (iii) $x^2 - 4x = 0$
   (iv) $2x^2 + 7x = 0$ (v) $x^2 - 2x = 8$ (vi) $4x^2 = 12x - 9$

4. Solve the following quadratic equations using the graphical method.
   (i) $x^2 + 7x - 8 = 0$ (ii) $2x^2 - x - 3 = 0$ (iii) $3 - 7x - x^2 = 0$
   (iv) $4 - 5x - x^2 = 0$ (v) $3x^2 + 5x + 2 = 0$ (vi) $2 - 4x - x^2 = 0$

1.22   Graphical Solutions to Linear-Quadratic Simultaneous Equations

Investigative Activity

1. Solve the simultaneous equations $y = x^2 - 3x + 2$ and $y = x - 1$.
2. Draw the graphs of $y = x^2 - 3x + 2$ and $y = x - 1$ on the same Cartesian plane.
3. State the coordinates of the points where the graphs intersect.
4. What conclusion do you draw?

We can solve simultaneous equations-one linear, one quadratic graphically by determining the point of intersection of the curve and the straight line in the same way as we solve simultaneous linear equations graphically.

Example

Use the graphical method to solve the simultaneous equations $y = x^2 - 5x + 4$ and $y = 2x - 6$.

Solution

Draw the line $y = 2x - 6$ on the same axes as the graph of $y = x^2 - 5x + 4$. The point of intersection of the two graphs, gives the solution. The table of values of $y = 2x - 6$ is as shown below. The figure below shows the two graphs.

x	2	3	4
y	3	4	2

The graphs intersect at $(2, -2)$ and $(5,4)$. Hence the solutions of the simultaneous equations are $x = 2$, when $y = -2$ and $x = 5$, when $y = 4$.

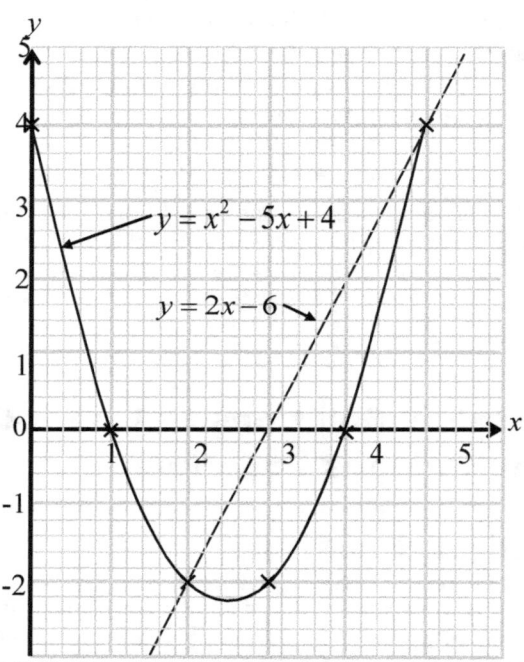

1.23   Quadratic Equations from the Graph of any Quadratic Function

It is possible to solve quadratic equations using the graph of any quadratic function. To do this, draw a suitable straight line on the same Cartesian plane as the graph of quadratic function. To obtain the equation of the suitable straight-line combine the quadratic equation and the quadratic function in such a way as to eliminate the squared terms.

 Example

1.   Use the graph of $f(x) = 4 - 3x - x^2$ above to solve the quadratic equation $x^2 + x - 6 = 0$.

Solution

The equation of the quadratic function is
$$f(x) = 4 - 3x - x^2 \text{ ... ... } \textcircled{1}$$

Rearranging $x^2 + x - 6 = 0$ we have
$$0 = 6 - x - x^2 \text{ ... ... } \textcircled{2}$$

Subtracting $\textcircled{2}$ from $\textcircled{1}$: $f(x) = -2 - 2x$.

Therefore the required straight line is $y = -2 - 2x$.

This straight line and the quadratic curve are shown in the graph below.

From the graph, the solution of the quadratic $x^2 + x - 6 = 0$ is $x = -3$ or $x = 2$.

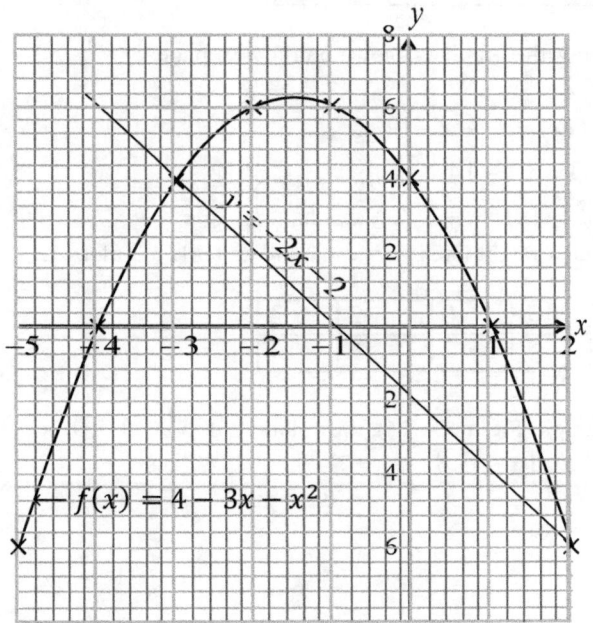

2. The function f is defined on the set $\mathbb{R}$ of real numbers as $f(x) = x^2 + x - 12$.
    (a) Using a scale of 1 cm for 2 units on the x-axis and 1 cm for 4 units on the y-axis, draw the graph of $y = f(x)$ for $x: -6 \le x \le 5$.
    (b) By drawing a suitable straight line on your graph, solve the equation $x^2 - 4 = 0$.
    (c) Also, solve from your graph the equation $2x^2 + 3x - 2 = 0$.

Solution

The table of values for the function is

x	$x^2 + x - 12$	$f(x)$
-6	$36 - 6 - 12$	18
-5	$25 - 5 - 12$	8

−4	$16 - 4 - 12$	0
−3	$9 - 3 - 12$	−6
−2	$4 - 2 - 12$	−10
−1	$1 - 1 - 12$	−12
0	$0 + 0 - 12$	−12
1	$1 + 1 - 12$	−10
2	$4 + 2 - 12$	−6
3	$9 + 3 - 12$	0
4	$16 + 4 - 12$	8
5	$25 + 5 - 12$	18

We now draw the graph as shown on the next page.

(b) $f(x) = x^2 + x - 12$ ....①
        $0 = x^2 - 14$ ....②
   ① − ② : $f(x) = x + 2$ ....③

The required straight line is $y = x + 2$ and its table of values is

x	0	−2	2
y	2	0	4

From the graph (Figure 34:14), the solution of $x^2 - 4 = 0$ is
$x \approx -3.8$ and $x \approx 3.8$.

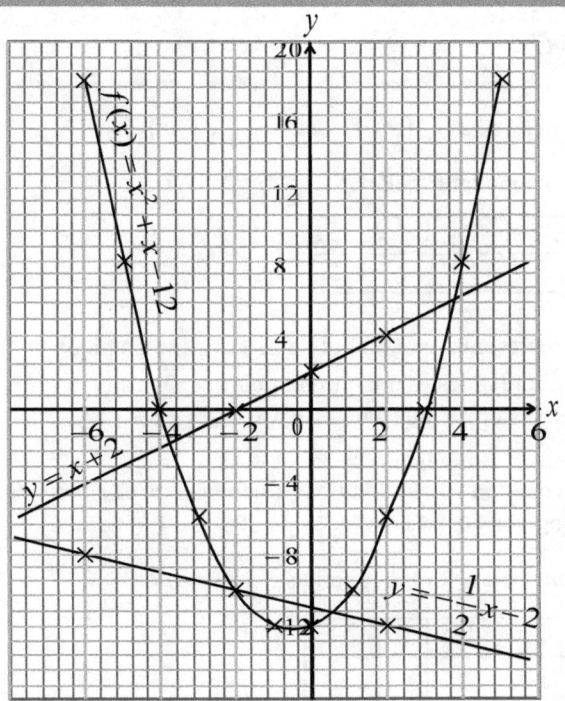

(c) We can solve the equation

$2x^2 + 3x - 2 = 0$ in a similar way.

$$f(x) = x^2 + x - 12 \dots\dots\dots\dots\dots\dots④$$

$$0 = 2x^2 + 3x - 2 \dots\dots\dots\dots⑤$$

$④ - \dfrac{1}{2} \times ⑤ : f(x) = -\dfrac{1}{2}x - 11$

The required straight line is the line $f(x) = -\dfrac{1}{2}x - 11$ whose table of values is

x	-6	2	6
y	-8	-12	-14

The line is shown in Figure 34:12 and from the graph, the solution of

$2x^2 + 3x - 2 = 0$ is $x \approx -2$ and $x \approx 0.5$.

29

Exercise 1:7

1. The function $f(x)$ is defined as $f(x) = \dfrac{1}{2}(x^2 + 2x - 5)$

   (a) Copy and complete the table below.

x	−5	−4	−3	−2	−1	0	1	2	3
$f(x)$									5

   (b) Taking 2 cm to represent 1 unit on both axes draw the graph of $y = f(x)$.
   (c) Write down the coordinates of the turning point.
   (d) Use your graph to estimate the value of x when $f(x) = 1.8$.

   (e) On the same axes, draw the line $y = 3x - 4$.
   (f) Write down the equation whose solutions are the points of intersection of the curve and the line $y = 3x - 4$.

2. (a) Complete the table below for the function $f(x) = 2x + \dfrac{1}{x}$.

x	0.25	0.5	1.0	1.5	2	2.5
$f(x)$	4.5					5.4

   (b) Using 2 cm to represent 1 unit on the y-axis and 4 cm to 1 unit on the x-axis draw the graph of $y = f(x)$.

   Use your graph to answer the following questions.
   (c) Write down the minimum value of $f(x)$ within the range of values 0.25 to 2.5.
   (d) Find the gradient of the curve at the point where $x = 0.5$
   (e) Solve for x, the equation $f(x) = 3.5$, giving your answer to 2 decimal places.

3. In an experiment involving two variables x and y, a student recorded the following readings:

x	−3	−2	−1	0	1	2	3	4	5
y	−6.1		0.3	2	2.7	2.4	1.1	−1.2	

   Given that the relationship between x and y is $2 + kx - \dfrac{1}{2}x^2$.
   (a) Find the value of k.
   (b) Determine the two missing values in the table.
   (c) Using 2 cm for 1 unit in the table, on both axes, draw the graph of this relationship.

   (d) Solve using your graph the equation $\dfrac{1}{2} + kx - \dfrac{1}{2}x^2 = 0$

   (e) Using your graph find the gradient of the curve at the point where $x = 0$.
4. Given the function $f(x) = 3x^2 - 8x - 7$.

(a) Construct a table of values of $f(x)$ for $-2 \le x \le 5$.

(b) Draw the graph of $y = f(x)$ for the range $-2 \le x \le 5$, taking 2 cm to represent 1 unit on the x-axis and 1 cm to represent 2 units on the y-axis. Use your graph to estimate the value of x for which:

(i) $3x^2 - 8x - 7 = 0$ (ii) $3x^2 - 8x - 7 = 19$

(iii) The gradient at $x = 3$.

5. Given that $f(x) = 2x^2 - 7x - 2$

(a) Copy and complete the following table

x	-2	-1	0	1	2	3	4	5
$f(x)$	20		-2					

(b) Taking 1 cm to represent 1 unit on the x-axis and 1 cm to represent 2 units on the y-axis, draw your graph to find
(i) The roots of the equation $f(x) = 0$.
(ii) The minimum value of $f(x)$ and the corresponding value of x.
(iii) The range of values of x for which $f(x) < 0$.

(c) By drawing a suitable straight line on your graph, estimate the roots of the equation $2x^2 - 8x + 1 = 0$.

1.24 Curves and Tangents

? **Brainstorming Exercise**

Given that a straight line is a tangent to a curve.
1. In how many points does the straight line touch the curve?
2. How many roots will you obtain if you solve the equation of the curve and the straight line simultaneously?

For a straight line to be a tangent to a curve at a point, the line must touch the curve at one and only one point. The implication of this is that equal roots will result when the equation of the curve and that of the straight line are solved simultaneously.

 Example

Determine which of the following lines, is a tangent to the curve $y = x^2$.

(a) $y = x + 2$ (b) $y = 6x - 9$

Solution

(a)
$$x^2 = x + 2$$
$$x^2 - x - 2 = 0$$
$$(x-2)(x+1) = 0$$
$$\Rightarrow x = 2 \text{ or } x = -1$$

Therefore, the line cuts the curve at two points.
Hence $y = x + 2$ is not a tangent to the curve.

(b)
$$x^2 = 6x - 9$$
$$x^2 - 6x + 9 = 0$$
$$(x-3)^2 = 0 \Rightarrow x = 3$$

Therefore, the line touches the curve at the point where $x = 3$. Hence, $y = 6x - 9$ is a tangent to the curve $y = x^2$.

1.25   Gradient of a Curve at a Point on the Curve

The gradient of a straight line is constant, but the gradient of a curve varies from point to point. The gradient of a curve at a particular point is the gradient of the tangent drawn to the curve at that point.

To find the gradient of a curve at a point,
(i) Draw the graph of the function,
(ii) Draw the tangent to the curve at that point.
(iii) Find the gradient of this tangent by choosing two points on the curve.

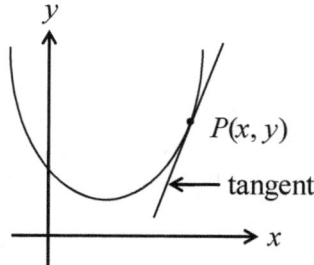

In finding the gradient of a function at a point using the graph of the function, the result depends on the accuracy of both the curve and the tangent, so take care to draw both the curve and the tangent.

Example

Find the gradient of the curve $y = x^2$ at the point $P(2,4)$.

Solution

The table of values and the graph of $y = x^2$ are as shown below.

x	-3	-2	-1	0	1	2	3
y	9	4	1	0	1	4	9

The graph is shown below.

Gradient m at P $= \dfrac{y_2 - y_1}{x_2 - x_1} = \dfrac{8 - 0}{3 - 1} = 4$

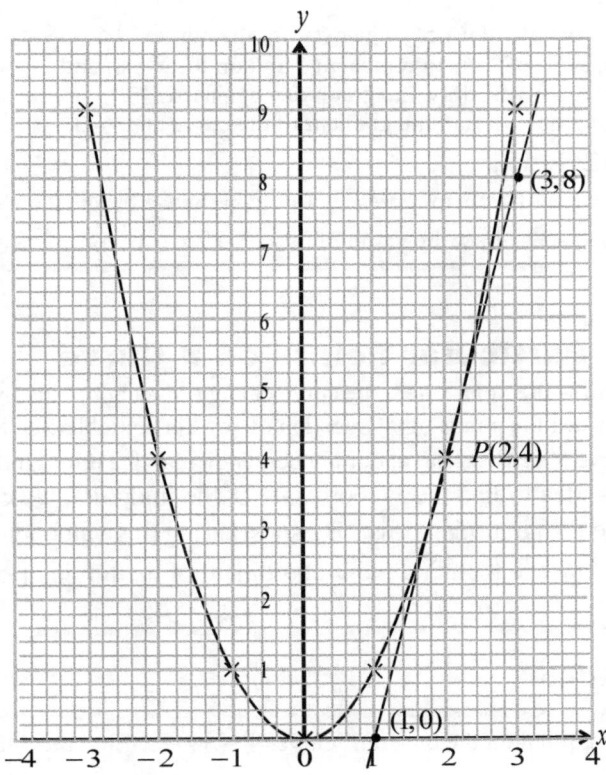

Exercise 1:8

1. By taking values of x in the range $-4 \le x \le 4$ and 1 cm to represent 1 unit on both axes, use a graphical method, to find the gradient of the curve
   $y = 2x^2 + x - 15$ at the point where $x = 1$.

2. If $y = 2x^2 + 3x - 2$,
   (a) Copy and complete the following table.

x	-4	-3	-2	-1	0	1	2	3
y		13						19

   (b) Using a scale of 1 cm to represent 1 unit on the x-axis and 1 cm to represent 2 units on the y-axis, draw graph $y = 2x^2 + 3x - 2$.
   (c) From your graph, find the gradient of the curve at the point where $x = 2$.

3. Given that $y = 2x^2 - 9x - 1$,
   (a) Make a table of values of y against x for the range $-1 \le x \le 6$.
   (b) Using a scale of 2 cm to represent 1 unit on the x-axis and 2 cm to represent 5 units on the y-axis, draw the graph of y against x.
   (c) Use your graph, to find the gradient of the curve at $x = 3$.

4. If $y = 5 - 2x - x^2$,
   (a) Using a scale of 2 cm to represent 1 unit on the x-axis and 2 cm to represent 2 units on the y-axis, draw the graph of $y = 5 - 2x - x^2$ for $-4 \le x \le 3$.

1.26 Particular Points on a Curve

The **turning point** is the point where the curve changes direction. The turning point is the midpoint between the points of intersection of any horizontal line and the curve since the quadratic curve is symmetrical about the perpendicular line through the turning point. We can use this idea to find the x-coordinate of the midpoint, which is equally the x-coordinate of the turning point. By substituting this value into the equation $y = f(x)$, we then find the y-coordinate of the turning point.

Example

Find the turning point on the quadratic curve $4 - 3x - x^2$ and the point where the curve cuts the y-axis.

Solution

The intercepts with the x-axis occur when, $f(x) = 0$ i.e. $4 - 3x - x^2 = 0$

$$(4+x)(1-x)=0$$
$$\Rightarrow \quad x=-4 \text{ or } x=1$$

But midpoint $\frac{x_1+x_2}{2} = \frac{-4+1}{2} = -\frac{3}{2}$

Therefore, the graph is symmetrical about the line $-\frac{3}{2}$.

Substitute in $f(x) = 4 - 3x - x^2$.

$$f\left(-\frac{3}{2}\right) = 4 - 3\left(-\frac{3}{2}\right) - \left(-\frac{3}{2}\right)^2 = \frac{25}{4}$$

Therefore, the turning point is $\left(-\frac{3}{2}, \frac{25}{4}\right)$.

The intercept with the y-axis occurs when $x = 0$.
$\Rightarrow f(0) = 4$ is the intercept with the y-axis.

1.27 Maximum and Minimum Values

 Investigative Activity

Examine all the quadratic curves you have been drawing in this topic and their respective functions and use them to answer the following questions.
1. How is the shape of a quadratic curve?
2. When does a quadratic curve have a maximum or a minimum?
3. Through what point on a quadratic curve must the axis of symmetry of the curve pass?
4. Given a quadratic function $y = ax^2 + bx + c$. What will be the value of the y-coordinate of the point the curve cuts the y-axis?

From the quadratic curves drawn above, we can see that;
1. A graph of quadratic function has either an ∩ shape or a U shape. The graph has a maximum point if it has an ∩ shape and a minimum point if it has a U shape. A maximum point exist if the coefficient of x^2 is less than zero (i.e. negative) whereas a minimum point exist if the coefficient of x^2 is greater than zero (i.e. positive).
2. A graph of quadratic function is symmetrical about an axis through the turning point.
3. The intercept with the y-axis is c.

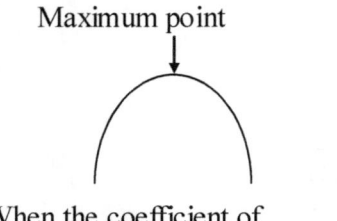

When the coefficient of
x^2 is less than zero
i.e. $a < 0$

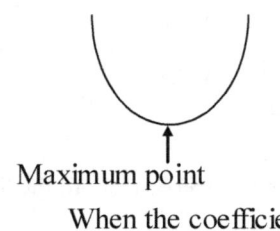

When the coefficient of
x^2 is greater than zero
i.e. $a > 0$

1.28 Nature of Roots of Quadratic Equations

There are six different types of graphs of quadratic functions depending on the discriminant ($\Delta = b^2 - 4ac$) and the value of the coefficient a of x^2.

For real and distinct roots, the graph cuts the x-axis at two points

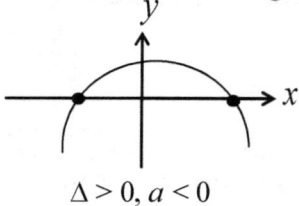

$\Delta > 0, a < 0$

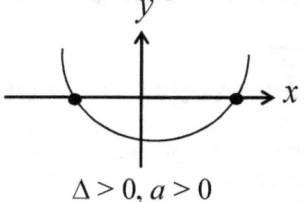

$\Delta > 0, a > 0$

For real and equal roots, the graph touches the x-axis at one and only one point.

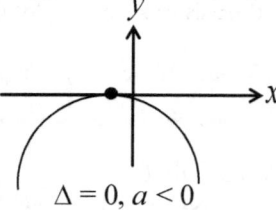

$\Delta = 0, a < 0$

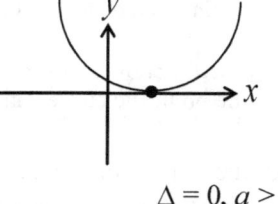

$\Delta = 0, a > 0$

For imaginary roots, the graph will neither cut nor touch the x-axis

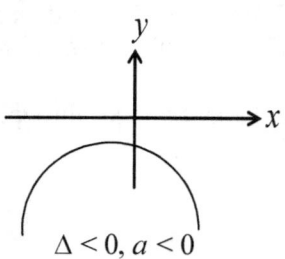

$\Delta < 0, a < 0$

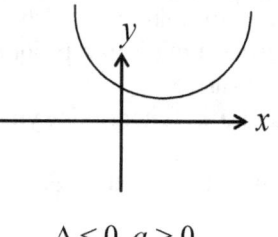

$\Delta < 0, a > 0$

Exercise 1:9

For each of the following quadratic functions and without plotting the graph of the function
(a) State the intercept of the curve with the y-axis.
(b) Find the intercepts of the curve with the x-axis.
(c) Determine the turning point on the curve and say whether it is a maximum or minimum point.

1. $f(x) = 5 + 4x - x^2$,   2. $f(x) = 3x^2 - 5x - 2$.

1.29 Speed time Graphs

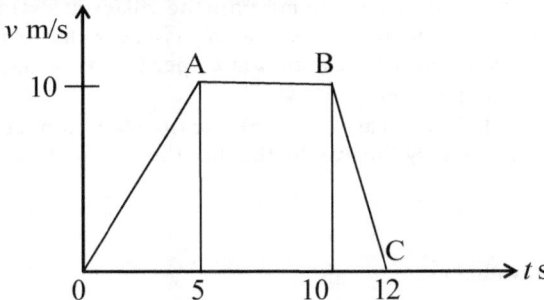

The figure above, shows a speed time graph for a body that starts from rest (i.e. with zero speed) at time $t = 0$ s and accelerates uniformly until it attains a velocity of 10 m/s at A. The body then moves with uniform speed of 10 m/s for 5 s from A to B. At B, the speed of the body decreases steadily until the body comes to rest again at $t = 12$ s. From this graph the total distance covered by the body is the area under the graph. Thus

Distance = Area of trapezium

Example

From the figure above, calculate the total distance covered by the body.

Solution

Total distance = Area of trapezium $= \dfrac{1}{2}(OC + AB)h$

$\Rightarrow$ Total distance $= \dfrac{1}{2}(12 + 5)10 = 85$ m

1.30 Distance Time Graphs

Distance-time graphs are very useful especially in determining the time or positions at which moving bodies meet. We usually plot the distance on the y-axis against the time on the x-axis. The graph of each body does not necessarily start from $(0,0)$ as the bodies may start at different distances from the origin and at different times. On a distance time graph, the gradient of the line or curve at any point gives the speed of the moving object.

 Example

A boy left a town at 8 a.m. trekking to his village 8 km away at a speed of 5 km/h. After trekking 2 km, he met a car coming from the village at a steady speed. Arriving in town at 8:30 a.m. the driver loaded passengers for 15 minutes then immediately returned to the village at the same speed. Draw a graph for these journeys and use it to determine
(a) The time the car left the village.   (b) The time the car overtook the man.
(c) The distance covered by the man by the time the car overtakes him.

Solution

The graph is shown below

From the graph,
(a) The car left the village at 8:45 a.m.
(b) The car overtakes the man at 8:48 a.m.
(c) The distance covered by the man 4.6 km.

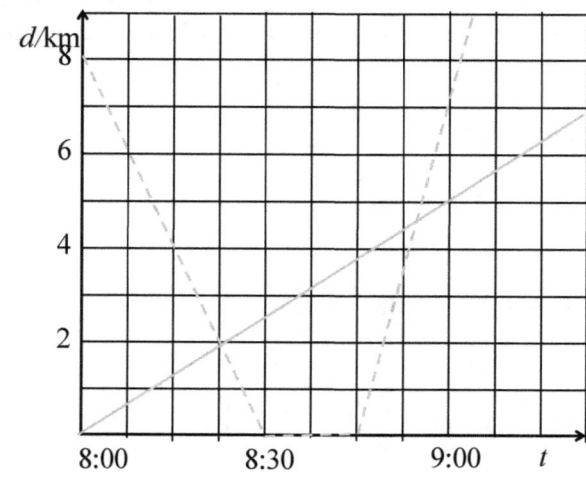

Exercise 1:10

1. A car starts from rest and is accelerated uniformly at the rate of 2 m/s for 6 s. It then maintains a constant speed for half a minute. The brakes are then applied and the vehicle uniformly retarded to rest in 5 s. Find the maximum speed reached in km/h and the total distance covered in metres.

2. Joan left school at 10.00 a.m. for her home 8 km away walking at the rate of 6 km per hour. She spent half an hour at home and a bike gave her a ride back to school traveling at 16 km per hour.
   (a) Determine the time she arrived their house.
   (b) Determine the time when she took off from the house to go back to school.
   (c) Determine the time she got back to school.
   (d) Draw the travel graph of her journey.
   (e) Using your graph or otherwise, determine how far she was away from the school at 12.00 noon. Express your answer to 1 decimal place.

Multiple Choice Exercise 1

1. A triangle with vertices at the points with coordinates $(-4,4)$, $(4,4)$, $(1,-1)$ is:
   [A] Right–angled   [B] Equilateral   [C] Isosceles   [D] Scalene

2. The statement which is true about the points $P(-1,-4)$, $Q(6,-5)$, $R(-1,5)$ and $S(3,2)$ is:
   [A] R and Q are in the second and third quadrants respectively.
   [B] P and S are in the fourth and third quadrants respectively.
   [C] S and R are in the first and second quadrants respectively.
   [D] P and Q are in the second and fourth quadrants respectively.

3. The coordinates of the midpoint of $P(-4,5)$ and $Q(2,1)$ are:
   [A] $(-1,3)$ [B] $(-1,-3)$ [C] $(1,-3)$ [D] $(1,3)$

4. A straight line passes through the points $(5,3)$ and $(8,4)$. The line:
   [A] is parallel to the x-axis   [B] slopes from left to right
   [C] is parallel to the x-axis   [D] slopes from right to left

5. Among the following, the straight line, which slopes from left to right, is the line, which passes through the points:
   [A] $(-1,5)$ and $(2,0)$ [B] $(3,-5)$ and $(3,3)$
   [C] $(-2,1)$ and $(1,1)$ [D] $(0,0)$ and $(3,6)$

6. Among the following, the straight line, which is parallel to the x-axis, is the line that passes through the points:
   [A] $(-1,5)$ and $(2,0)$ [B] $(3,-5)$ and$(3,3)$
   [C] $(-2,1)$ and $(1,1)$ [D] $(0,0)$ and $(3,6)$

7. Among the following, the straight line, which slopes from right to left, is the line, which passes through the points:
   [A] $(-2,5)$ and $(-2,3)$ [B] $(3,-5)$ and $(8,-5)$
   [C] $(-2,1)$ and $(1,1)$ [D] $(0,0)$ and $(3,6)$

8. The point, which lies on the line $y = 2x-5$ is:
    [A] (1,3) [B] (2,5) [C] (3,−1) [D] (3,1)

9. The gradient of the line $2x+3y = 12$ is:
    [A] $-\dfrac{2}{3}$ [B] $-\dfrac{3}{2}$ [C] $\dfrac{2}{3}$ [D] $\dfrac{3}{2}$

10. The intercept of the line $2x+3y = 12$ with the x-axis is:
    [A] 2   [B] 3   [C] −6   [D] 6

11. The intercept of the line $2x+3y = 12$ with the y−axis is:
    [A] 2   [B] 3   [C] 4   [D] −4

12. The area of the triangle OAB, where O is the origin and A and B are the points where the line cuts the x- and y-axes respectively is:
    [A] 12 un^2 [B] 24 un^2 [C] 6 un^2 [D] 18 un^2

13. The gradient of the line $2x+y = 8$ is:
    [A] 4   [B] 2   [C] −2   [D] −8

14. Given that the lines $2x+y = 8$ and $6y-mx = 3$ are parallel. The value of m is:
    [A] 2   [B] −2   [C] $\dfrac{1}{2}$ [D] $-\dfrac{1}{2}$

15. Given that the lines $2x+y = 8$ and $6y-mx = 3$ are perpendicular. The value of m is:
    [A] 2   [B] −2   [C] $\dfrac{1}{2}$ [D] $-\dfrac{1}{2}$

16. In the form $y = mx + c$, the equation of the line $2x-3y+5 = 0$ is:
    [A] $y = -\dfrac{2}{3}x + \dfrac{5}{3}$ [B] $y = \dfrac{2}{3}x - \dfrac{5}{3}$
    [C] $y = \dfrac{2}{3}x + \dfrac{5}{3}$ [D] $y = -\dfrac{2}{3}x - \dfrac{5}{3}$

17. The gradient of the line $2x-3y+5 = 0$ is:
    [A] $-\dfrac{5}{3}$ [B] $\dfrac{5}{3}$ [C] $-\dfrac{2}{3}$ [D] $\dfrac{2}{3}$

18. The intercept of the line $2x-3y + 5 = 0$ with the x−axis is:
    [A] $-\dfrac{5}{2}$ [B] $\dfrac{5}{2}$ [C] $-\dfrac{5}{3}$ [D] $\dfrac{5}{3}$

19. The coordinates of the midpoint of (3,2) and (−1,0) are:
    [A] (−1,−1)   [B] (1,1)   [C] (−1,1)   [D](1,−1)

20. The gradient of the straight line which passes through the points (−1,0) and (0,−2) is:
    [A] −2   [B] 2   [C] $-\dfrac{1}{2}$ [D] $\dfrac{1}{2}$

21. The distance between the points (3,2) and (0,−2) is:
    [A] 3   [B] $\sqrt{3}$ [C] 5   [D] $\sqrt{5}$

22. The coordinate of the point of intersection of the lines $x-y = 3$ and $x+2y = 6$ is:

[A] (4,1)   [B] (6,1)   [C] $\left(\dfrac{9}{4},\dfrac{3}{4}\right)$ [D] (1,4)

23. Given the straight lines
$L_1 : 2x = 3y + 5$, $L_2 : 2y = 4x + 3$, $L_3 : 2x + 4y = 5$, $L_4 : 3x - 2y = 3$

The lines, which are parallel, are:
[A] L_1 and L_4 [B] L_2 and L_4 [C] L_2 and L_3 [D] L_1 and L_4

24. Given the straight lines
$L_1 : 2x = 3y + 5$, $L_2 : 2y = 4x + 3$, $L_3 : 2x + 4y = 5$, $L_4 : 3x - 2y = 3$

The lines are perpendicular are:
[A] L_1 and L_4 [B] L_2 and L_4 [C] L_2 and L_3 [D] L_1 and L_4

25. Given the straight lines
$L_1 : 2x = 3y + 5$, $L_2 : 2y = 4x + 3$, $L_3 : 2x + 4y = 5$, $L_4 : 3x - 2y = 3$

The line, which passes through the point $(-1,-3)$ is:
[A] L_1 [B] L_2 [C] L_3 [D] L_4

26. P is the midpoint of the line segment joining the points $A(2,-3)$ and $B(4,5)$. C is the point $(5,9)$. The equation of the line PC is:
[A] $y = -4x + 11$ [B] $y = 4x - 11$
[C] $y = 4x + 11$ [D] $y = -4x - 11$

27. Given the straight lines
$L_1 : y - 2x = 5$ $L_2 : y = 2x + 3$
$L_3 : 4y = -2x - 6$ $L_4 : 2y = x - 5$

The lines, which are parallel, are:
[A] L_1 and L_2 [B] L_2 and L_3 [B] [C] L_1 and L_4 [D] L_3 and L_4

28. Given the straight lines
$L_1 : y - 2x = 5$ $L_2 : y = 2x + 3$
$L_3 : 4y = -2x - 6$ $L_4 : 2y = x - 5$

One of the pair of perpendicular lines are:
[A] L_1 and L_2 [B] L_2 and L_3 [C] L_1 and L_4 [D] L_3 and L_4

29. Given that,
$L_1 : y = 2x + 3$ $L_2 : 2x + y = 5$
$L_3 : 4y = 2x + 3$ $L_4 : x + \dfrac{y}{-2} = 1$

The lines, which are parallel, are:
[A] L_1 and L_4 [B] L_2 and L_4 [C] L_2 and L_3 [D] L_1 and L_4

30. Given that,
$L_1 : y = 2x + 3$ $L_2 : 2x + y = 5$
$L_3 : 4y = 2x + 3$ $L_4 : x + \dfrac{y}{-2} = 1$

The lines are perpendicular are:
[A] L_1 and L_4 [B] L_2 and L_4 [C] L_2 and L_3 [D] L_1 and L_2

31. The point of intersection of the lines $y - 2x = 5$ and $3y = 4x + 9$ is:
[A] $(3,-1)$ [B] $(3,1)$ [C] $(-3,1)$ [D] $(-3,-1)$

32. The line $3x - 5y = 15$ cuts the y-axis at:

[A] (0,5)　　　　　　[B] (0,–3)　　　　　[C] (5,0)　　　[D] (–3,0)

33. The line $3x–5y = 15$ cuts the x-axis at:

[A] (0,5)　　　　[B] (0,–3)　　　　　[C] (5,0)　　　　　[D] (–3,0)

34. The gradient of the line $3x–5y = 15$ is

[A] $-\dfrac{5}{3}$　[B] $\dfrac{5}{3}$　[C] $-\dfrac{3}{5}$　[D] $\dfrac{3}{5}$

35. The value of m for which the lines $y = 2x–13$ and $y – mx = –3$ are parallel is:

[A] -2　[B] 2　[C] $\dfrac{1}{2}$　[D] $-\dfrac{1}{2}$

36. The value of m for which the lines $y = 2x–13$ and $y – mx = –3$ are perpendicular is:

[A] -2　[B] 2　[C] $\dfrac{1}{2}$　[D] $-\dfrac{1}{2}$

37. Given that the lines $y = 2x–13$ and $y–mx = –3$ are perpendicular. Their point of intersection is:

[A] (–4,–5)　　　　　[B] (4,5)　　　　　[C] (–4,5)　　　　　[D] (4,–5)

3. The gradient of the lines of the line $x+2y = 2$ is:

[A] $-\dfrac{1}{2}$　[B] $\dfrac{1}{2}$　[C] 2　[D] -2

38. The intercept of the line $x+2y = 2$ with the x-axis is:

[A] (0,–1)　[B] (0,1)　[C] (2,0)　[D] (–2,0)

39. The intercept of the line $x+2y = 2$ with the y-axis is:

[A] (0,–1)　　　　　[B] (0,1)　　　　[C] (2,0)　　　　　[D] (–2,0)

40. The area of the triangle between the line $x+2y = 2$ and the coordinate axes is:

[A] 1 un^2　　　　　[B] 2 un^2　　　[C] 5 un^2　　　　　[D] $\sqrt{5} \text{ un}^2$

41. $P(3,0)$ and $Q(5,2)$ are two points on a straight line. The equation of the line PQ is:

[A] $x + y = 3$　　　[B] $y = x–3$　　　[C] $y = x+3$　　　　[D] $y = –x –3$

42. Given the lines

$L_1: y = 2x–4,$　　$L_2: 2y+x–6 = 0,$　　$L_3: y = \dfrac{1}{3}x + 7$　　$L_4: 3y = x–5.$

The two perpendicular lines are:

[A] L_1 and L_3　　[B] L_2 and L_4　[C] L_3 and L_4　　[D] L_1 and L_2

43. Given the lines

$L_1: y = 2x–4,$　　$L_2: 2y+x–6 = 0,$　　$L_3: y = \dfrac{1}{3}x + 7$　　$L_4: 3y = x–5.$

The two parallel lines are:

[A] L_1 and L_3　　[B] L_2 and L_4　[C] L_3 and L_4　　[D] L_1 and L_2

44. Given that the points $P(2,2)$, $Q(3,5)$ and $R(4,a)$ are collinear, the coordinates of R are:

[A] (4,–8)　　　[B] (3,–7)　　　[C] (4,7)　　　[D] (4,8)

45. The y intercept on the line $6x+3y–7 = 0$ is:

[A] $\dfrac{7}{6}$ [B] −7   [C] $\dfrac{7}{3}$ [D] −2

46. The x intercept on the line $6x+3y-7 = 0$ is:

[A] $\dfrac{7}{6}$ [B] −7   [C] $\dfrac{7}{3}$ [D] −2

47. The gradient of the line $6x+3y-7 = 0$ is:

[A] $\dfrac{7}{6}$ [B] −7   [C] $\dfrac{7}{3}$ [D] −2

48. The straight line parallel to the y-axis passes through the points:
[A] (−1,5) and (2,0)   [B] (3,−5) and (3,3)
[C] (−2,1) and (1,1)   [D] (0,0) and (3,6)

49. P is the midpoint of the line segment joining the point $A(2,-3)$ and $B(4,5)$. C is the point (5,9). The gradient of the line PC is:
[A] −4   [B] 4   [C] 11   [D] −11

50. P is the midpoint of the line segment joining the point $A(2,-3)$ and $B(4,5)$. C is the point (5,9). The intercept of the line PC with the y-axis is:

[A] 11   [B] −11   [C] $\dfrac{11}{4}$ [D] $-\dfrac{11}{4}$

51. P is the midpoint of the line segment joining the point $A(2,-3)$ and $B(4,5)$. C is the point (5,9). The intercept of the line PC with the x-axis is:

[A] 11   [B] −11   [C] $\dfrac{11}{4}$ [D] $-\dfrac{11}{4}$

52. In the figure below, the intercept of the line l_1 with the y-axis is:

[A] $-\dfrac{5}{4}$ [B] $-\dfrac{4}{5}$ [C] 4   [D] −5

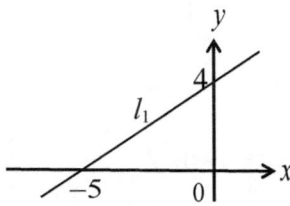

53. In the figure above, the intercept of the line l_1 with the x-axis is:

[A] $-\dfrac{5}{4}$ [B] $-\dfrac{4}{5}$ [C] 4   [D] −5

54. In the figure above, the gradient of the line l_1 is:

[A] $-\dfrac{5}{4}$ [B] $-\dfrac{4}{5}$ [C] −4   [D] −5

55. In the figure above, the equation of the line l_1 is:
[A] $5y = 4x - 20$ [B] $4y = 5x - 20$
[C] $5y = -4x + 20$ [D] $4y = -5x + 20$

56. In the figure below, the intercept of the line l_2 with the y-axis is:

[A] $-\dfrac{3}{5}$ [B] $\dfrac{3}{5}$ [C] 6 [D] 10

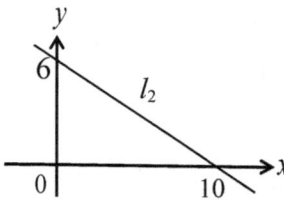

57. In the figure above, the intercept of the line l_2 with the x-axis is:

[A] $-\dfrac{3}{5}$ [B] $\dfrac{3}{5}$ [C] 6 [D] 10

58. In the figure above, the gradient of the line l_2 is:

[A] $-\dfrac{3}{5}$ [B] $\dfrac{3}{5}$ [C] 6 [D] 10

59. In the figure above, the equation of the line l_2 is:

[A] $5y = 3x + 30$ [B] $3y = 5x - 30$ [C] $5y = 3x - 30$ [D] $3y = -5x + 30$

60. The root of the equation represented by the graph below is:

[A] 4 [B] 7 [C] −4 [D] −7

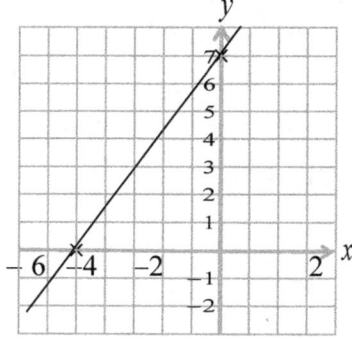

61. The graph of $f(x) = x^2 + x - 6$ is most likely:

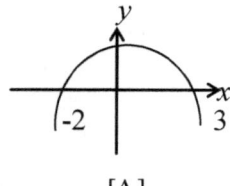

[A]

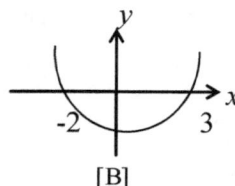

[B]

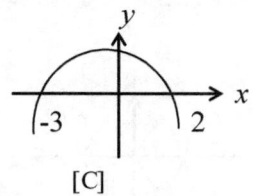

[C]

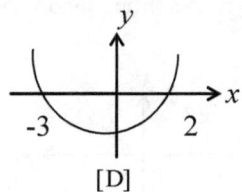

[D]

62. The graph of $y = 4 - 3x - x^2$ is most certainly:

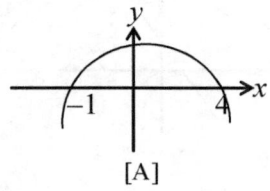

[A]

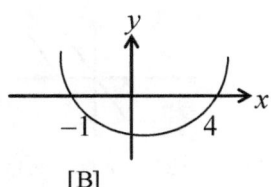

[B]

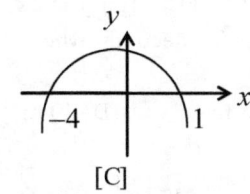

[C]

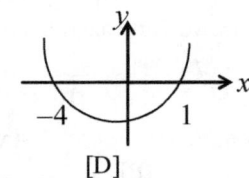

[D]

63. The equation, which represents the sketch graph in the figure (a) below is:

[A] $y = 8 - 2x + x^2$ [B] $y = 8 + 2x + x^2$ [C] $y = 8 - 2x - x^2$ [D] $y = 8 + 2x - x^2$

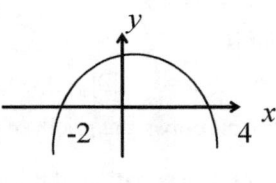

(a)

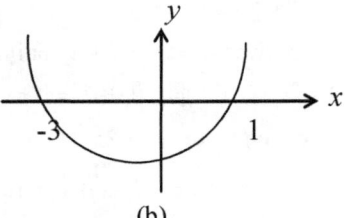

(b)

64. The equation, which represents the sketch graph in figure (b) above is:

[A] $y = x^2 - 2x - 3$ [B] $y = x^2 + 2x - 3$ [C] $y = x^2 + 2x + 3$ [D] $y = x^2 - 2x + 3$

65. The figure below is the graph of the function:

[A] $y = x^2 - x - 6$ [B] $y = x^2 + x - 6$ [C] $y = x^2 + x + 6$ [D] $y = x^2 - x + 6$

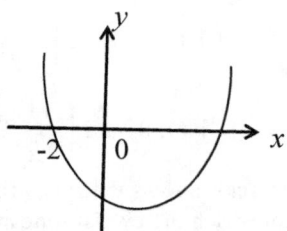

66. The graph representing inconsistent lines is:

[A]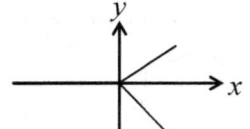

[B]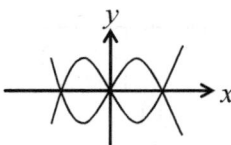

[C]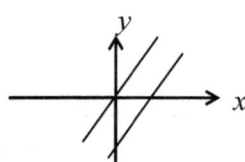

[D]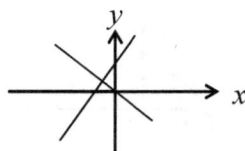

67. The line, which is a tangent to the curve, $y = x^2$ is:

    [A] $y = 2x - 3$ [B] $y = 2x - 2$ [C] $y = 2x - 1$ [D] $y = 2x$

68. A stone thrown vertically upward moves s metres in t seconds, where

    $s = 80t - 5t^2$. The maximum height it reaches is:

    [A] 640 m   [B] 320 m   [C] 75 m   [D] 40 m

69. The maximum value of $x^2 - 4x + 5$ is:

    [A] −5   [B] 0   [C] −1   [D] 1

70. The function whose curve has a maximum point is:

    [A] $y = 8 - 2x - x^2$ [B] $y = 8 + 2x + x^2$

    [C] $y = x^2 + 2x + 8$ [D] $y = x^2 - 2x + 8$

71. The function whose curve has a minimum point is:

    [A] $f(x) = 12 + 8x - x^2$ [B] $f(x) = 12 - 8x + x^2$ [C] $f(x) = 12 - 8x - x^2$ [D] $f(x) = -12 + 8x - x^2$

72. To solve the equation $y = 2x^2 - 5x - 1$ a student draws the graph of the

    curve $y = 2x^2 - 5x - 1$ and a straight line PQ. The equation of the straight

    line PQ is:

    [A] $y = 1$ [B] $y = 0$ [C] $y = 3$ [D] $y = -3$

73. The equation of a curve is $y = 2x^2 - x - 1$. The intercept with the y-axis is:

    [A] $(-1, 0)$ [B] $(1, 0)$ [C] $(0, -1)$ [D] $(0, 1)$

74. The equation of a curve is $y = 2x^2 - x - 1$. The intercepts with the x-axis is:

    [A] $(0, 1)$ and $\left(0, -\dfrac{1}{2}\right)$ [B] $(-1, 0)$ and $\left(-\dfrac{1}{2}, 0\right)$

    [C] $(-1, 0)$ and $\left(\dfrac{1}{2}, 0\right)$ [D] $(1, 0)$ and $\left(-\dfrac{1}{2}, 0\right)$

75. The figure below not drawn to scale shows the speed time graph of a cyclist.
    The length of time in hours for which the cyclist rode at uniform speed is:
    [A] 3   [B] 4   [C] 6   [D] 18

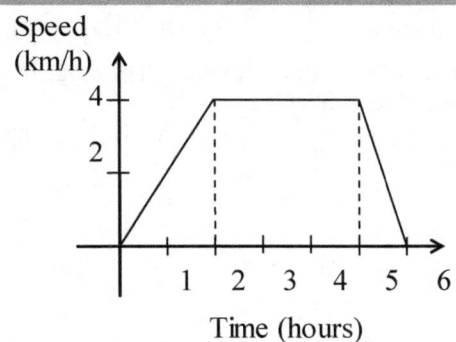

Speed (km/h)

Time (hours)

76. A train of length 100 m is traveling at a speed of 40 km per hour. In minutes, the length of time taken for the train to pass completely over a bridge of length 0.7 km is:

[A] 1.2 minutes   [B] 0.02 minutes   [C] 12 minutes   [D] 0.2 minutes

77. A particle P moves so that its distance, s metres, from a fixed point O, at time t seconds is given by $s = 20 + 24t - t^2$. The distance traveled during the first 3 seconds of motion is:

[A] 20 m   [B] 72 m   [C] 83 m   [D] 92 m

78. The table of values for $y = x - 6$ is:

[A]
x	−5	−8	−7
y	1	−14	−13

[B]
x	−5	−8	−7
y	−11	−2	−13

[C]
x	−5	−8	−7
y	−11	−14	−13

[D]
x	−5	−8	−7
y	1	−2	−1

79. The equation, which corresponds to the table of values, is:

[A] $y = 4 + 5x$ [B] $y = 3 + 6x$ [C] $y = 5 + 4x$ [D] $y = 6 + 3x$

Input (x)	1	2	3	4	5
Output (y)	9	12	15	18	21

80. Figure (a) below shows the graph of a quadratic function. The range of values of the discriminant Δ and the coefficient of x^2 is:

[A] $\Delta > 0, a > 0$ [B] $\Delta < 0, a < 0$ [C] $\Delta > 0, a < 0$ [D] $\Delta < 0, a > 0$

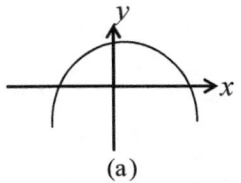

(a)   (b)

81. Figure (b) above shows the graph of a quadratic function. The range of values of the discriminant Δ and the coefficient of x^2 is:

47

[A] $\Delta>0, a>0$ [B] $\Delta<0, a<0$ [C] $\Delta>0, a<0$ [D] $\Delta<0, a>0$

82. The figure below shows the graph of a quadratic function. The range of values of the discriminant Δ and the coefficient of x^2 is:

[A] $\Delta>0, a>0$ [B] $\Delta<0, a<0$ [C] $\Delta>0, a<0$ [D] $\Delta<0, a>0$

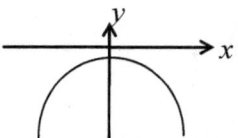

Topic 2

EUCLIDEAN GEOMETRY

Objectives

At the end of this topic, the learner should be able to:

1. Observe, describe, recognize and identify a plane figure
2. Make sketches of plane figures.
3. State the properties of triangles.
4. State the properties of quadrilaterals.
5. Calculate the perimeter and area of plane figures.

2.1 Review and Revision of Plane Figures

We did much work on plane figures in modules 2 and 6. The learner may go back and revise book 1, topics 11 and book 2, topic 8 before continuing.

The following is an extract from modules 2 and 6 of the formulae for finding the perimeter and area of some popular plane figures.

Name and Diagram	Perimeter	Area
Triangle a h c b	$P = a + b + c$	$A = \dfrac{1}{2} bh$ $A = \sqrt{p(p-a)(p-b)(p-c)}$
Square l l	$P = 4l$	$A = l^2$
Rectangle l w	$P = (l + w) \times 2$	$A = lw$
Parallelogram a h b	$P = (a + b) \times 2$	$A = bh$
Rhombus x l y	$P = 4l$	$A = \dfrac{1}{2} xy$

Trapezium		
a, h, b		$A = \dfrac{1}{2}(a + b)h$
Circle		
r	$C = 2\pi r$	$A = \pi r^2$

Recall the Pythagoras theorem which states that in a right angled triangle;

$$c^2 = a^2 + b^2$$
$$a^2 = c^2 - b^2$$
$$b^2 = c^2 - a^2$$

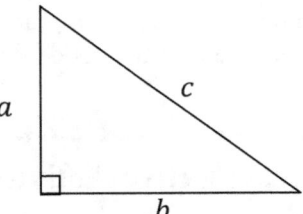

Review and Revision Exercise

(1) State the number of sides and draw a diagram of each of the following polygons.
    (a) an octagon   (b) a triangle   (c) a heptagon
    (d) a quadrilateral   (e) a hexagon   (f) a pentagon

(2) Draw a diagram of each of the following quadrilaterals.
    (a) a square   (b) a trapezium   (c) a rhombus
    (d) a Kite   (e) a parallelogram   (f) a rectangle

(3) Which of the quadrilaterals in (2) above is a parallelogram? Give reasons for your answer.

(4) Which of the quadrilaterals in (2) above is not a parallelogram? Give reasons for your answer.

(5) State the major differences between

(a) a square and a rectangle   (b) a Kite and a rhombus

(c) a square and a rhombus

(6) Draw a diagram of each of the following triangles.

(a) an equilateral triangle   (b) a right angled triangle

(c) an obtuse angled triangle   (d) an isosceles triangle

(e) an acute angled triangle   (f) a scalene triangle

2.2   Angle Properties of Polygons

Sum of Interior Angles of a Triangle

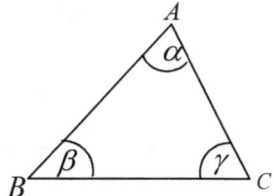

In the figure above, the angles α, β and γ shown are called the **interior angles** of the triangle, because they are inside the triangle.

 ## Investigative Activity

(1) Draw a triangle ABC (any size) and produce AC to D as shown in figure (a) below.

(2) Cut out the vertices A and B of the triangle ABC and arrange at C as in figure (b) below.

(3) Do they fit exactly onto the angle θ the exterior angle of the triangle? This shows that the exterior angle of a triangle is equal to the sum of the two interior opposite angles.

(4) Do the three angles α, β and γ lie and fit exactly on the straight line AD?

(5) Use your answer in (4) to deduce the numerical value of $\alpha + \beta + \gamma$?

(6) Make a general statement concerning the interior angles of a triangle.

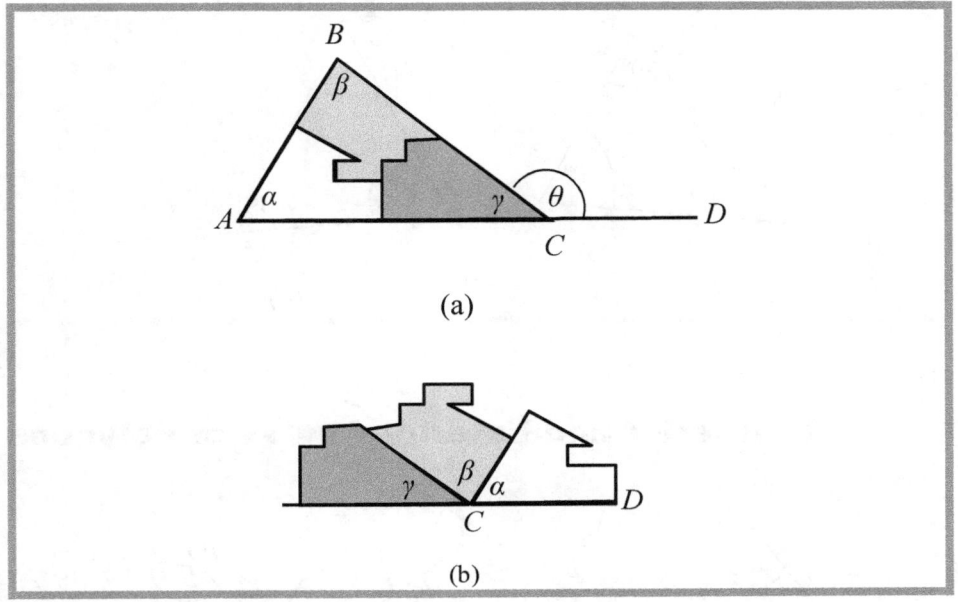

(a)

(b)

Since the angles α and β fit into the angle θ and all the three angles lie on a straight line, the above investigation reveals that

(1) The exterior angle of a triangle is equal to the sum of the two interior opposite angles. $\theta = \alpha + \beta$
(2) The sum of the interior angles of a triangle is $180°$ (or two right angles). In other word, $\alpha + \beta + \gamma = 180°$

Exercise 2:1

Find the value of the lettered angles in figure (a) to (d) below.

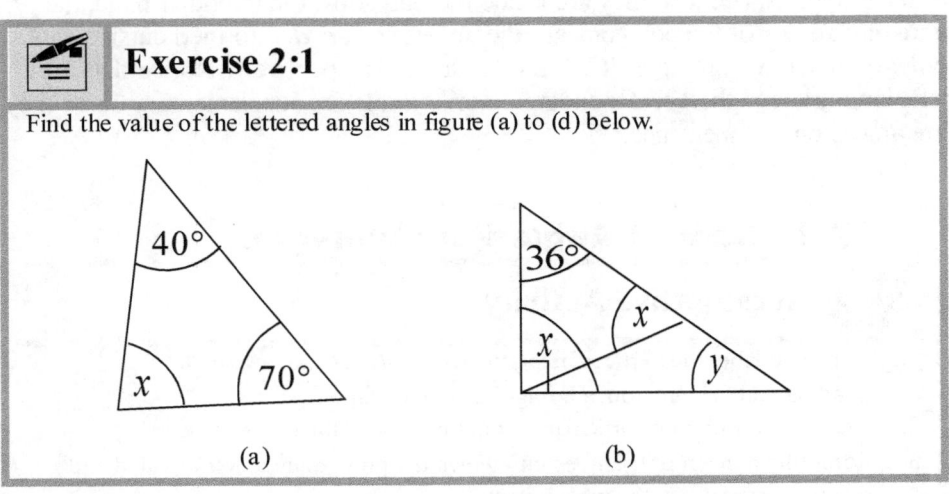

(a)

(b)

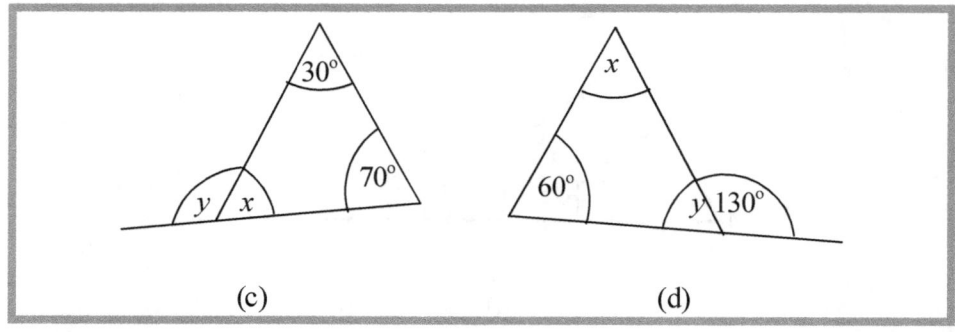

(c)   (d)

2.3   Interior and Exterior Angles of Polygons

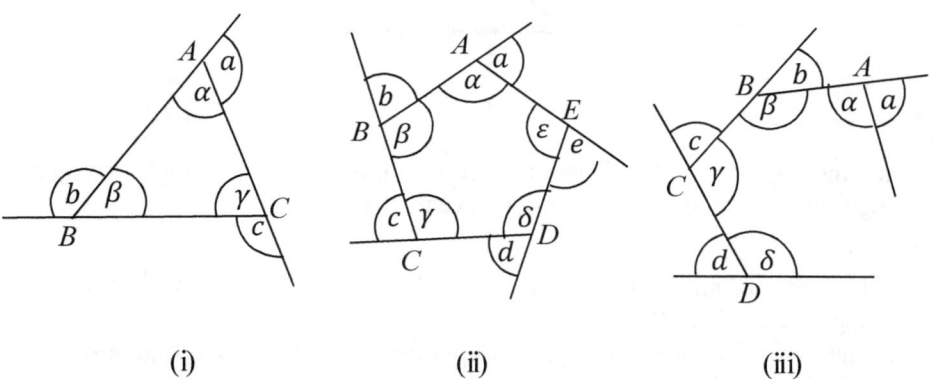

(i)   (ii)   (iii)

In the above figure, the angles α, β, γ, δ, ε, shown are called the **interior angles** of the polygons, because they are inside the polygons. On the other hand, the **exterior angles** of the polygons are the angles a, b, c, d, e formed outside each polygon when we produce AC, BA, CB, DC, ED respectively. Notice that each interior angle and the corresponding exterior angle are supplementary since they are angles on a straight line.

2.4   Sum of Angles of a Polygon

 Investigative Activity

1. Draw and label the vertices using the letters A, B, C...of a triangle, a quadrilateral, a pentagon, a hexagon and a heptagon
2. Draw lines from one particular vertex say A to all the other vertices.
3. Count the number of triangles into which the polygon has been divided and record your result in the table below.

Use the following example which has been done for an octagon to help you.

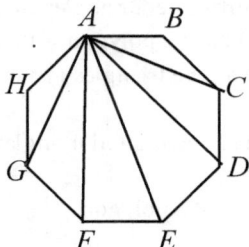

No of sides of polygon	No of triangles insides polygon	Sum of interior angles
3		
4		
5		
6		
7		
8	8	1080°

4. Study the results in the table carefully and;
   (i) Predict the number of triangles inside a polygon with 9 sides.
   (ii) Predict the sum of all the interior angles in a convex polygon with 9 sides.
5. Write down an expression for the number of triangles insides a polygon with n sides.
6. Write down an expression for the sum of interior angles of a polygon with n sides.
7. Deduce the sum of the exterior angles of a polygon.
8. Repeat instruction 1.
9. Draw lines from the vertex to the centre of each polygon as in the following figure.

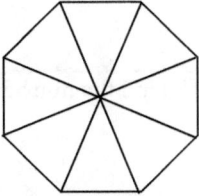

10. Count the number of triangles into which the polygon has been divided and record your result in the table below.
11. Study your results in the table carefully.
12. Predict the number of triangles inside a polygon with 9 sides.
13. What is the sum of the angles of all the triangles in a convex polygon with 9 sides?

14. What is the sum of the angles at the centre of the polygon?
15. Use (13) and (14) to find the sum of interior angles of a polygon with 9 sides.
16. What is the number of triangles inside a polygon with n sides?
17. What is the sum of the angles of all the triangles in a convex polygon with n sides?
18. Write down an expression for the sum of interior angles of a polygon with n sides.
19. Deduce the sum of exterior angles of a polygon.

No of sides of polygon	No of triangles insides polygon	Sum of interior angles
3		
4		
5		
6		
7		
8	8	$1080°$

The above investigative exercise, lead us into the polygon theorems that follow. The polygon theorems are actually extensions of Chasles' Theorem.

2.5 Polygon Theorems

In a convex polygon with n sides,

(1) The sum of the interior angles is $(n-2)180°$
(2) The sum of the exterior angles is 4 right angles (or $360°$) no matter the value of n.

 Example

1. Each angle of a regular polygon is $170°$. Find the number of sides of the polygon.

   Solution
   Each exterior angle $= 180° - 170° = 10°$
   Since sum of exterior angles is equal to $360°$, the number of sides must be $360° \div 10° = 36$.
   Therefore, the polygon has 36 sides.

2. One angle of a hexagon is $140°$ and 5 angles are equal. Find the value of each of the 5 angles.

Solution

The exterior angle corresponding to $140°$ is $180°-140° = 40°$

Therefore, the sum of the other 5 exterior angles is $360°-40° = 320°$

So the value of each of the 5 exterior angles $320° \div 5 = 64°$.

Therefore, the corresponding angle is $180°-64° = 116°$

Hence, the value of each of the 5 angles is $116°$.

Note!!

In solving problems on polygons it is often easier to use the theorem on sum of exterior angles of a polygon rather than that on the sum of interior angles, though both lead to the same answer.

 Exercise 2:2

1. Four angles of a hexagon are $130°$, $160°$, $112°$, and $140°$. If the remaining angles are equal, find the size of each.
2. Each of the exterior angles of a regular polygon is $100°$ less than the interior angle. Calculate the size of the exterior angle.
3. The sum of the interior angles of an n-sided convex polygon is double the sum of the exterior angles. Find the value of n.
4. The angles of a pentagon are $x°$, $2x°$, $(x+30)°$, $(x-10)°$, and $(x+40)°$. Find the value of x.
5. The sum of the angles of a polygon is $1800°$. Calculate the number of sides of the polygon
6. How many sides, has a convex polygon with each interior angle equal to $150°$?
7. How many sides, has a convex polygon with each exterior angle equal to $20°$?
8. Find the size of an interior angle of a regular ten-sided polygon.
9. Determine whether it is possible to have a regular convex polygon with exterior angles, (a) $20°$ (b) $16°$ (c) $15°$.
   If so, state the number of sides of the polygon.
10. One exterior angle of a polygon is $54°$, and other exterior angles are each $34°$. Calculate the number of sides of the polygon.
11. Four of the sides of a pentagon are $72°$, $100°$, $120°$, and $140°$ in that order. Find the remaining angle; hence prove that two of the sides are parallel.
12. $ABCDEFGH$ is a regular octagon. Calculate $\angle AFC$.
13. AB, BC, CD are three consecutive sides of a regular polygon. If $\angle ACB = 15°$, calculate the number of sides of the polygon and $\angle ACD$.
14. $ABCDEFGHI$ is a regular nonagon. Calculate
    (a) The angles ABC, CAE, ACG (b) The angles between AE and CG
15. P, Q, R, S are four adjacent vertices of a regular octagon. Calculate the angles PQR and RSP.
16. AB and BC are adjacent sides of a regular dodecagon. The perpendicular from C meets AB produced at D. Calculate $\angle BCD$.
17. By dividing an n-sided convex polygon into isosceles triangles each with one

vertex at the centre and a side of the polygon as its base, prove that:
(a) The sum of the interior angles of any n-sided convex polygon is $(2n-4)$ right angles or $(n-2)180°$.
(b) The sum of the exterior angles of any n-sided convex polygon is $360°$.

2.6   Symmetry

Wat Phra Kaeo

The Wat Phra Kaeo temple in Thailand built in 1782 is one of the greatest symmetrical architectural designs in the world.

Symmetry is the correspondence of parts on opposite sides of a point, line or plane. Symmetry is a very important phenomenon in disciplines such as architecture, mathematics, biology, physics, mineralogy etc. The bodies of many animals for instance exhibit bilateral symmetry on two opposite sides of a linear axis, or a median plane.

2.7   Types of Symmetry

Objects exhibit two main types of symmetry. Namely, mirror symmetry and rotational symmetry.

2.8   Orthogonal or Mirror Symmetry

The terms orthogonal symmetry, line symmetry, reflective symmetry and mirror symmetry are synonymous. Mirror symmetry is that property of an object in which one of the halves of the object appears like the other part when placed against a mirror. If a plane shape is folded along the mirror line, the two parts fit together.
In plane figures, the mirror line is called a **line of symmetry.**

Investigative Activity

Examine the following plane shapes. The dotted lines are the lines through which one half of the plane shape can be folded so that it exactly fits on the other half.

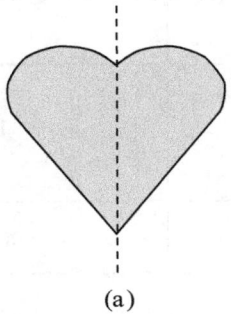

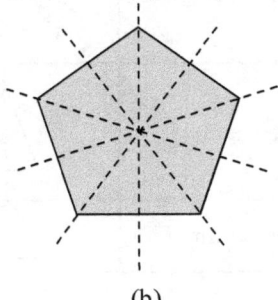

(a) (b)

1. On a square paper, draw each of the following shapes.
   (i) Isosceles trapezium (ii) Square (iii) Rhombus
   (iv) Rectangle (v) Kite (vi) Equilateral triangle
   (vii) Isosceles triangle (i) Parallelogram
2. Using a pair of scissors or a blade; cut out each of the shapes in (1) above.
3. By folding each shape, identify all the possible lines through which one half of each figure can be folded and super imposed on the other half.

The table below shows some plane figures and their lines of symmetry. The lines of symmetry are shown dotted.

	Plane Figure	Number of lines of symmetry
Isosceles trapezium		1
Square		4
Rhombus		2
Rectangle		2

Kite		1
Equilateral triangle		3
Isosceles triangle		1
Parallelogram		none

How many lines of symmetry does the four-pointed star below have? Can you count them? How many lines of symmetry does the regular octagon below have?

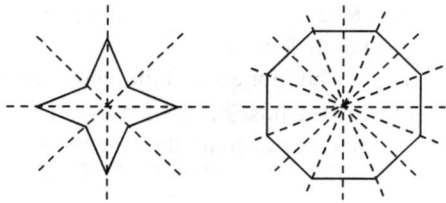

The four-pointed star has four lines of symmetry. The regular octagon has eight lines of symmetry.

2.9   Rotational or Radial symmetry

The terms radial symmetry, rotational symmetry or point symmetry are synonymous. **Radial symmetry** is the proportional arrangement of similar parts of a body around a central axis, as in the case of jellyfish or starfish. If we rotate a circle any amount about the centre, it remains unchanged. Therefore, a circle exhibits radial symmetry of infinite order about its centre.

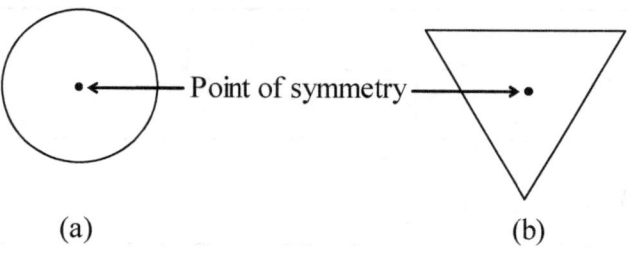

(a)   (b)

An equilateral triangle remains unchanged in three different positions when rotated about the point of symmetry. Hence, an equilateral triangle exhibits radial symmetry of order 3.

Any regular polygon with n sides has rotational symmetry of order n because when rotated, its shape is exactly repeated in n different positions as shown in the table below.

Regular Polygon	Order
Equilateral triangle	3
Square	4
Regular pentagon	5
Regular hexagon	6
n-gon	n

We earlier saw that a parallelogram has no line of symmetry. Figure (a) below shows that a parallelogram exhibits point symmetry about the point O. A_2, B_2, C_2, and D_2 can respectively fit on A_1, B_1, C_1 and D_1. In other words if the parallelogram is rotated through $180°$, it remains unchanged.

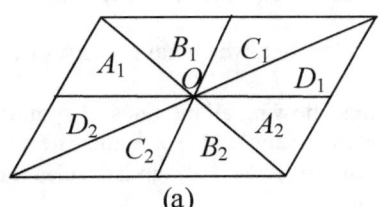

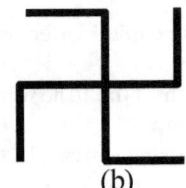

(a)   (b)

What is the order of rotational symmetry of swastika (the figure (b) above) the German army batch?

Exercise 2:3

1. How many lines of symmetry has
   (a) A regular hexagon   (b) a regular pentagon   (c) A regular octagon
2. By drawing and showing, using dotted lines state the number of axes (lines) of symmetry, if any which the following have:
   (a) A kite   (b) An equilateral triangle   (c) A square
   (d) A rectangle   (e) A regular pentagon   (f) A rhombus
   (g) A parallelogram   (h) A regular hexagon
3. Plot the following points on square paper and connect them.
   (a) A (2, 5), B (4, 12), C (6, 10)
   (b) A (4, 5), B (6, 5), C (6, 10), D (4, 10)
   (c) A (4, 5), B (6, 5), C (6, 10), D (4, 12)

(d) A (4,10), B (8,10), C (10,15), D (6,15)

(e) A (4, 3), B (6, 10), C (8, 3), D (10, 10)

(f) A (2, 2), B (6, 5), C (6, 15), D (2, 15)

(g) $A(0,10)$, $B(5,5)$, $C(10,5)$, $D(15,10)$, $E(15,15)$, $F(10,20)$, $G(5,20)$, $H(0,15)$

(h) A (0, 5), B (5, 0), C (0,-5), D (-5, 0)

(i) $A(0,2)$, $B(2,3)$, $C(6,3)$, $D(6,1)$, $E(2,1)$

List the shapes, which possess (i) Line symmetry (ii) Point symmetry

Write down the order of rotational symmetry for each of the shapes or write N if the shape does not possess rotational symmetry.

4. Complete each of the diagrams in Figure 45:14, so that each will have rotational symmetry of order
   (i) Two   (ii) three   (iii) four

5. Name all the capital letters of the English alphabet, which have one and only one axes of symmetry.

6. State the number of planes of symmetry which each of the following has.
   (a) A cube   (b) A cone   (c) A sphere   (d) A square pyramid.

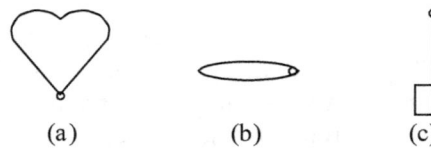

(a)   (b)   (c)

7. List all the capital letters of the English alphabet, which have at least two axes of symmetry.

8. List all the capital letters of the English alphabet, which have no axes of symmetry.

9. Draw each of the following plane figures showing all the lines of symmetry.
   (a) a square   (b) an equilateral triangle   (c) a regular hexagon
   (d) an isosceles triangle   (e) a parallelogram   (d) an isosceles trapezium

10. $\mathscr{E}$- capital letters of the English alphabet
    V = letters with a vertical line symmetry
    H = letters with a horizontal line symmetry
    R = letters which have rotational symmetry
    Find $H \cap V$. What conclusion do you draw?

✍ Multiple Choice Exercise 2

1. The statement, which is always true of a rhombus, is:
   [A] All the angles are complementary   [B] All the sides are equal
   [C] The adjacent angles are equal   [D] All the angles are equal

2. The plane shape, which is not a quadrilateral, is:
   [A] A kite   [B] a rhombus   [C] A pentagon   [D] a parallelogram

3. The plane shape, which is not an example of a quadrilateral, is:
   [A] Square   [B] Trapezium   [C] Rhombus   [D] Triangle

4. The statement, which is not true of a parallelogram, is:
   [A] It has more than 4 sides   [B] Opposite sides are parallel.

[C] It has exactly 4 sides.   [D] The sum of its angles is 360°.
5. The name of a seven sided plane figure is:
    [A] An octagon   [B] a pentagon   [C] A hexagon   [D] a heptagon
6. A polygon with all its interior angles less than 180° is definitely:
    [A] a convex polygon   [B] a regular polygon
    [C] a re-entrant polygon   [D] a quadrilateral
7. A triangle with vertices (−4, 4), (4,4) ,(0,−1) is:
    [A] Right-angled   [B] equilateral   [C] Isosceles   [D] scalene
8. The largest angle of any triangle:
    [A] Must always be an acute angle.   [B] Can sometimes be an acute angle.
    [C] Can never be a right-angle.   [D] Must always be an obtuse angle.
9. A quadrilateral with one pair of sides equal is:
    [A] A rhombus   [B] a parallelogram   [C] A rectangle   [D] a trapezium
10. The assertion about a rhombus, which may not be true, is:
    [A] The opposite angles are equal.
    [B] The diagonals bisect the angles through which they pass.
    [C] The diagonals are equal.
    [D] Opposite sides are equal.
11. A quadrilateral whose diagonals bisect at right angles is:
    [A] A rectangle   [B] a parallelogram   [C] A trapezium   [D] a rhombus
12. The property/properties, which do not characterize a rectangle, is/are:
    I.   The diagonals bisect at right angles
    II. Opposite sides are equal and parallel
    III. Each of its angles is a right angle
    [A] I only   [B] II only   [C] III only   [D] II and III only
13. In the figure below, the value of the angle marked y is:
    [A] 28° [B] 62° [C] 118° [D] 152°

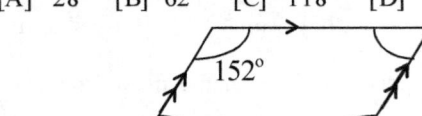

14. The value of the angle marked x in the figure below is:
    [A] 140° [B] 130° [C] 110° [D] 70°

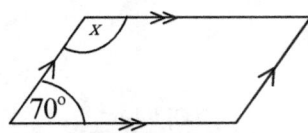

15. The value of the angle marked y in the figure below is:
    [A] 270° [B] 210° [C] 190° [D] 95°

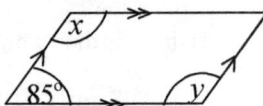

16. In the figure above, the sum of x and y is:
    [A] 270° [B] 210° [C] 190° [D] 95°
17. The values of x, y, and z in the figure below are respectively:

[A] $130°, 50°, 130°$ [B] $140°, 40°, 140°$
[C] $150°, 30°, 150°$ [D] $120°, 60°, 120°$

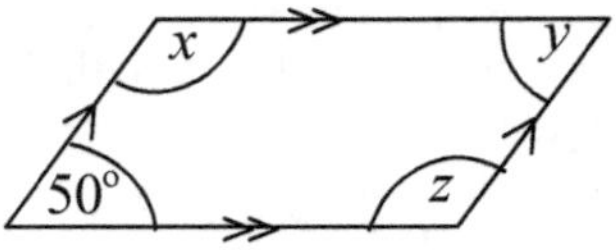

18. The name given to the figure below is:

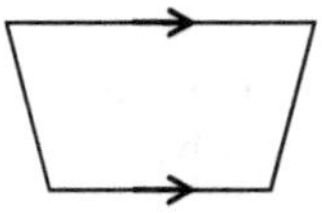

[A] A parallelogram   [B] a trapezium   [C] A rhombus   [D] a rectangle

19. Let $\mathscr{E} = \{x : x \text{ is a polygon}\}$
$A = \{x : x \text{ is a regular polygon}\}$
$B = \{x : x \text{ is a quadrilateral}\}$
Then an element of $A \cap B$ is:
[A] A square   [B] a rhombus   [C] A trapezium   [D] a rectangle

20. Let $\mathscr{E} = \{x : x \text{ is a polygon}\}$
$A = \{x : x \text{ is a regular polygon}\}$
$B = \{x : x \text{ is a quadrilateral}\}$
Then an element of $A' \cap B$ can never be:
[A] A square   [B] a trapezium   [C] A rhombus   [D] a rectangle

21. The sum of the angles of a square is:
[A] $90°$ [B] $120°$ [C] $180°$ [D] $360°$

22. If the perimeter of a square is 36 cm then the area of the square in square centimetres is:
[A] 81 cm^2 [B] 36 cm^2 [C] 9 cm^2 [D] 36^2 cm^2

23. The area of a square is $x^2 \text{ cm}^2$. Its perimeter is:

24. The figure below is a rectangle. If the perimeter is 36 m, the area of the rectangle is:
[A] x^4 cm   [B] $4x^2$ cm   [C] $4x$ cm   [D] $2x$ cm

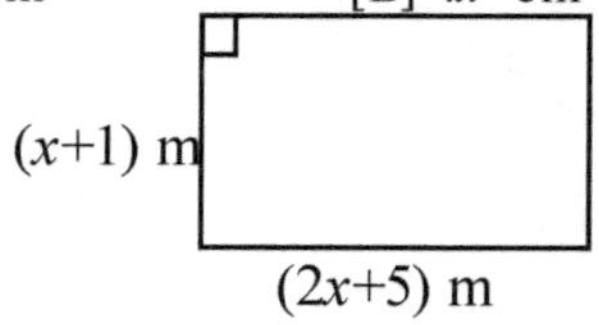

$(x+1)$ m

$(2x+5)$ m

[A] 64 m^2 [B] 65 m^2 [C] 84 m^2 [D] 124 m^2

25. One side of a rectangular field is 8 m and the diagonal is 10 m. The area of the rectangle is:
[A] 80 m^2 [B] 36 m^2 [C] 40 m^2 [D] 48 m^2

26. The length of a rectangle is twice the width. If the length is 8 cm, the perimeter in centimetres is:
[A] 24 cm   [B] 32 cm   [C] 48 cm   [D] 12 cm

27. The area of a rectangle with width 4 m and diagonal 8 m is:
[A] $8\sqrt{3} \text{ m}^2$ [B] $12\sqrt{3} \text{ m}^2$ [C] $16\sqrt{3} \text{ m}^2$ [D] 48 m^2

28. A woman pastes a rectangular photograph 15 cm by 9 cm on a rectangular card

and leaves a margin of 2.5 cm round the photograph, the perimeter of the card is:
[A] 58 cm   [B] 68 cm   [C] 98 cm   [D] 228 cm

29. In a trapezium, the lengths of the parallel sides are 4 cm and 6 cm and the perpendicular distance between these sides is 3 cm. The area of the trapezium is:
[A] 36 cm^2 [B] 18 cm^2 [C] 30 cm^2 [D] 15 cm^2

30. The area in square units of the trapezium in the figure below is:
[A] 24 un^2 [B] 40 un^2 [C] 32 un^2 [D] 30 un^2

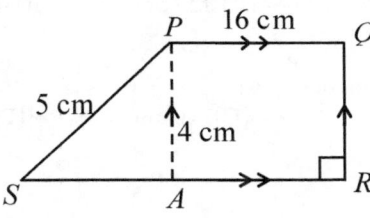

31. The perimeter of the trapezium *PQRS* in the figure below is:

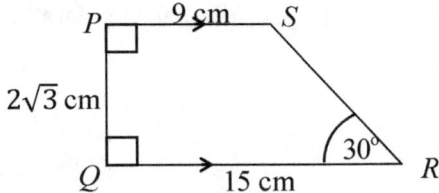

[A] 24 cm   [B] 44 cm   [C] 36 cm   [D] 70 cm

32. The area of the parallelogram in the figure below is:
[A] 2736 cm^2 [B] 936 cm^2 [C] 1368 cm^2 [D] 1872 cm^2

33. The lengths of the parallel sides of a trapezium are 5 cm and 7 cm. If its area is 120 cm^2, the perpendicular distance between the parallel sides is:
[A] 5.0 cm   [B] 6.9 cm   [C] 20.0 cm   [D] 10.0 cm

34. In the figure below, *PQRS* is a trapezium in which $|PS| = 9$ cm, $|QR| = 15$ cm, $|PQ| = 2\sqrt{3}$ cm, $\angle PQR = 90°$ and $\angle QRS = 30°$. The area of the trapezium is:
[A] $24\sqrt{3}$ cm^2 [B] $36\sqrt{3}$ cm^2 [C] $42\sqrt{3}$ cm^2 [D] $72\sqrt{3}$ cm^2

35. In the figure below, *ABCD* is a parallelogram, in which *DE* is perpendicular to *BC*, *DE* = 5 cm, *BC* = 25 cm and *AD* = 45°. The size of *ABC* is:
[A] 45°   [B] 90°   [C] 135°   [D] 145°

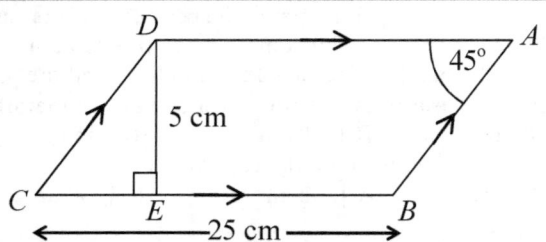

36. In the figure above, *ABCD* is a parallelogram, in which *DE* is perpendicular to *BC*, *DE* = 5 cm, *BC* = 25 cm and *AD* = 45°. The length of *EC* is:
    [A] 25 cm   [B] 20 cm   [C] 15 cm   [D] 5 cm

37. In Figure 31:56, *ABCD* is a parallelogram, in which *DE* is perpendicular to *BC*, *DE* = 5 cm, *BC* = 25 cm and *AD* = 45°. The area of *ABCD* is:

    [A] 625 cm^2 [B] 250 cm^2 [C] 125 cm^2 [D] $62\frac{1}{2}$ cm^2

38. The area of triangle *PQR* in the figure below is:
    [A] 24 cm^2 [B] 12 cm^2 [C] 10 cm^2 [D] 48 cm^2

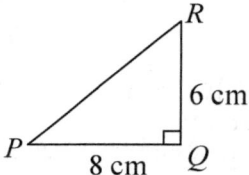

39. The area of a triangle, the sides of which are 5 cm, 4 cm and 3 cm long, is:
    [A] 30 cm^2 [B] 15 cm^2 [C] 12 cm^2 [D] 6 cm^2

40. The area of an equilateral triangle of side 16 cm is:
    [A] $64\sqrt{3}$ cm^2 [B] $32\sqrt{3}$ cm^2 [C] 96 cm^2 [D] 128 cm^2

41. The diagonals *AC* and *BD* of a rhombus *ABCD* are 16 cm and 12 cm respectively. The area of the rhombus must be:
    [A] 36 cm^2 [B] 48 cm^2 [C] 60 cm^2 [D] 96 cm^2

42. The area of a rhombus is 24 cm^2 and one of its diagonals is 8 cm. The side of the rhombus is:
    [A] 4.3 cm   [B] 5 cm   [C] 6 cm   [D 10 cm

43. By the number of sides the polygon in the figure below is:
    [A] An octagon   [B] a pentagon   [C] A quadrilateral   [D] a hexagon

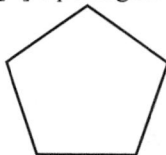

44. The polygon, which has the shape of the Cameroon flag, is:
    [A] An octagon   [B] a pentagon   [C] A hexagon   [D] a quadrilateral

45. A shape that has two more sides than a football field is:

[A] Pentagonal   [B] octagonal   [C] Hexagonal   [D] decagonal

46. A Polygon has 4 more sides than a rectangle. The polygon is:
    [A] Pentagon   [B] decagon   [C] Hexagon   [D] octagon

47. A seven sided plane figure is called:
    [A] an octagon [B] a pentagon [C] a hexagon [D] a heptagon

48. A polygon with all its interior angles less than $180°$ is definitely:
    [A] a convex polygon   [B] a regular polygon
    [C] a re-entrant polygon   [D] a quadrilateral

49. The polygon which has the shape of the Cameroon flag is:
    [A] octagon   [B] pentagon   [C] hexagon   [D] quadrilateral

50. The number of sides in a regular polygon whose interior angle is $135°$ is:
    [A] 7   [B] 8   [C] 10   [D] 12

51. The number of sides in a regular polygon with each of its interior angles equal to $108°$ is:
    [A] 4   [B] 5   [C] 6   [D] 7

52. The number of sides in a regular polygon with each of its interior angles equal to $140°$ is:
    [A] 7   [B] 8   [C] 9   [D] 10

53. The number of sides in a regular polygon with each of its interior angles $120°$ is:
    [A] 4   [B] 6   [C] 7   [D] 8

54. The number of sides in a regular polygon with one interior angle $160°$ is:
    [A] 10   [B] 36   [C] 18   [D] 20

55. In the figure below, *WXYZ* is a rhombus and $\angle WYZ=20°$. The value of angle *XZY* is:
    [A] $20°$ [B] $30°$ [C] $60°$ [D] $70°$

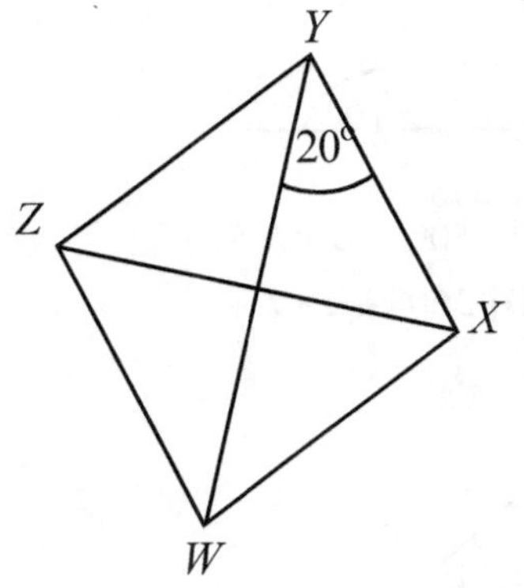

56. The number sides in a regular polygon with one interior angle $160°$ is:
    [A] 10   [B] 36   [C] 18   [D] 20

57. The real name of the plane the figure below and some of its possible names are:
    [A] parallelogram; quadrilateral, rhombus.
    [B] Rhombus; quadrilateral, trapezium.
    [C] Trapezium; quadrilateral, polygon.
    [D] Quadrilateral, parallelogram, polygon.

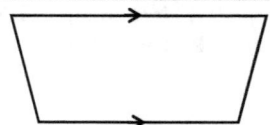

58. The best name of the quadrilateral, which has 4 congruent sides, is:
    [A] Parallelogram   [B] rhombus   [C] Trapezoid   [D] rectangle
59. A triangle has sides 3, 5, 8 and angles 25°, 85°, 70°. By the measure of its angles and its sides the triangle is:
    [A] Isosceles, obtuse   [C] isosceles, acute
    [B] Scalene, obtuse   [D] scalene, acute
60. The diagram in the figure below which is a polygon is:

    [A]   [B]   [C]   [D]

61. The angles of a pentagon are $x°$, $2x°$, $(x+60)°$, $(x+10)°$, $x°$, $(x-10)°$.
    The value of x is:
    [A] 80°   [B] 75°   [C] 60°   [D] 40°

62. In the figure below, the true relation is:
    [A] $a + b + x = 180°$ [B] $a = b + x$
    [C] $a - b = 180° - x$ [D] $a + b = x + 180°$

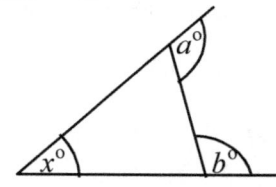

63. In the figure below, x is equal to:
    [A] $a + b + c$ [B] $360° - (a + b + c)$
    [C] $a + b + c + 180°$ [D] $360° - a + b + c$

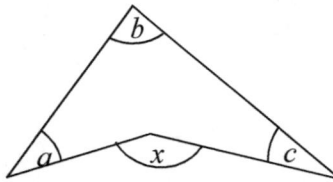

64. In the figure below, the size of angle ACB is:
    [A] 40°   [B] 50°   [C] 60°   [D] 80°

65. In the figure below, $|PQ| = |PR| = RS$ and $\angle RPS = 32°$. The value of $\angle QPR$ is:

[A] 64°   [B] 52°   [C] 32°   [D] 26°

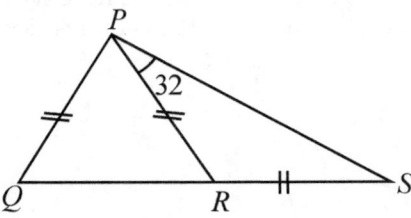

66. In the figure below, ABC is a triangle, BC is produced to D, $|AB| = |AC|$, $\angle BAC = 50°$. The value of $\angle ACD$ is:

[A] 115°   [B] 65°   [C] 60°   [D] 50°

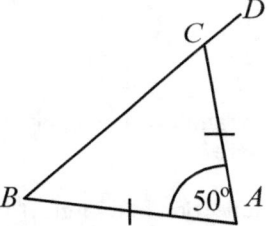

67. We can subdivide a regular polygon centre O into isosceles triangles identical to triangle POQ below. The number of such triangles in the polygon is:

[A] 12   [B] 10   [C] 9   [D] 8

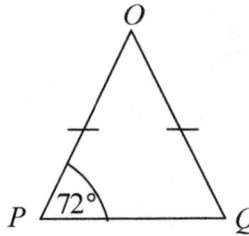

68. The value of angle t in the figure below is:

[A] 115°   [B] 120°   [C] 125°   [D] 145°

69. The property, which makes a rhombus different from every other parallelogram, is:

[A] All sides are equal.   [B] Opposite sides are parallel.
[C] Opposite angles are equal.   [D] Diagonals bisect each other at right angles.

70. The property, which makes a square a unique rectangle, is:

[A] All sides are equal.   [B] Opposite sides are parallel.
[C] Opposite angles are equal.   [D] Diagonals bisect each other at right angles.

71. The property, which makes a square a unique rhombus, is:

[A] All sides are equal.   [B] Opposite sides are parallel.
[C] Opposite angles are equal.   [D] Diagonals bisect each other at right angles.

72. Among the properties, the property, which makes a rectangle a special parallelogram, is:
    [A] Opposite sides are equal. [B] Opposite sides are parallel.
    [C] Diagonals bisect each other. [D] Diagonals are equal in length.

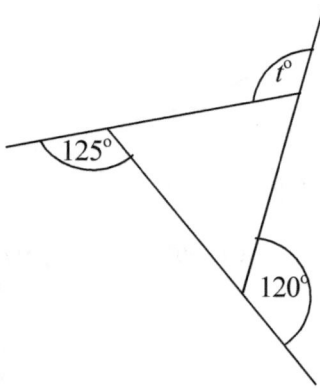

73. The order of rotational symmetry of a pyramid with a square base and axis through the vertex and centre of the square is:
    [A] 2   [B] 3   [C] 4   [D] 5

74. The graph in figure (b) which shows the reflection of triangle *ABC* in Figure (a) in the line *XY* is:

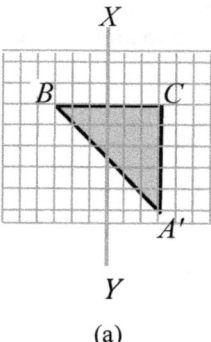

(a)

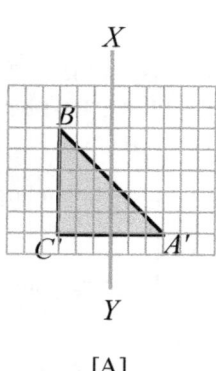

[A]

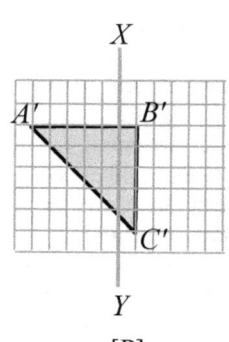

[B]

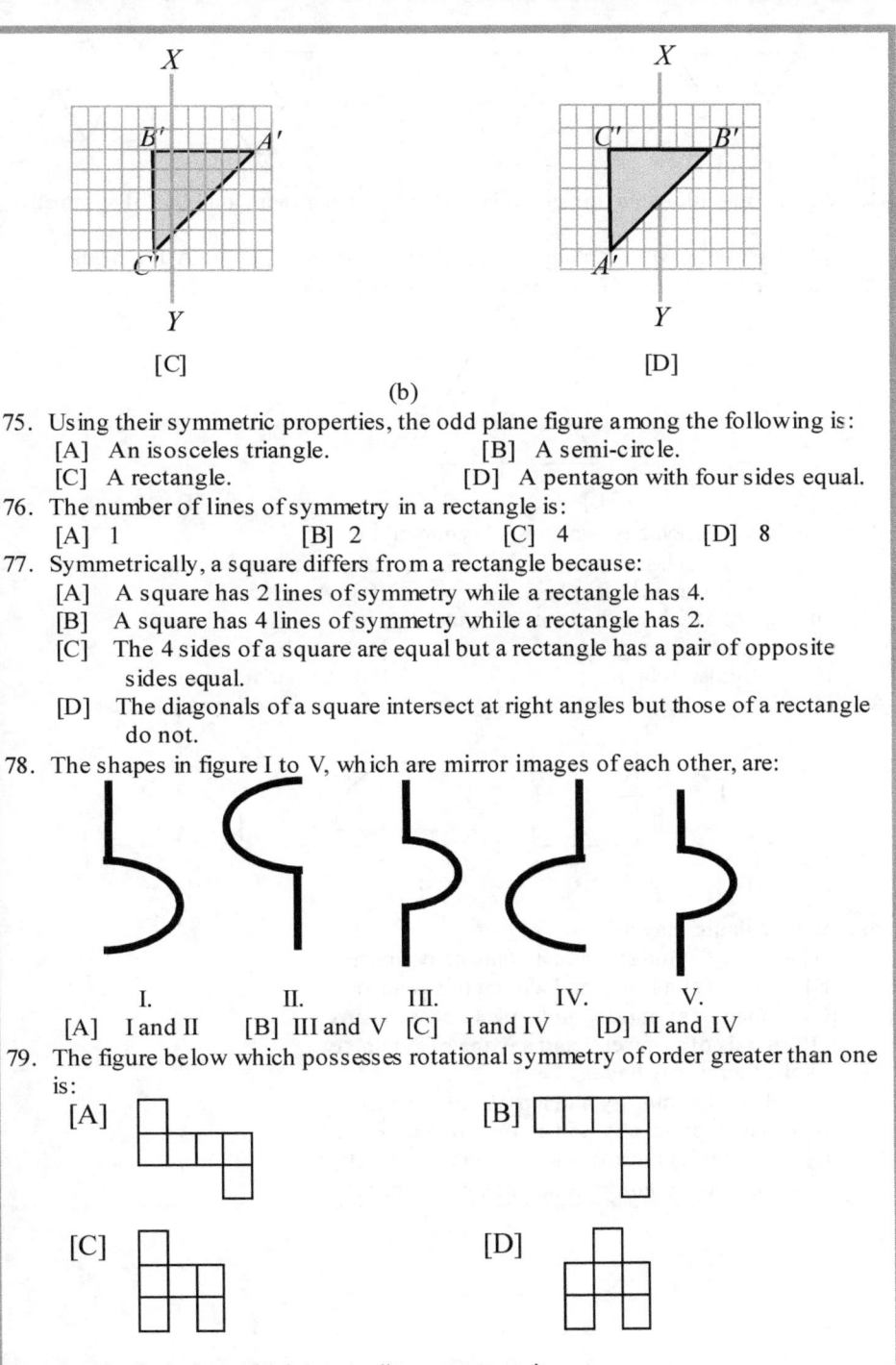

[C]   [D]

(b)

75. Using their symmetric properties, the odd plane figure among the following is:
   [A]   An isosceles triangle.   [B]   A semi-circle.
   [C]   A rectangle.   [D]   A pentagon with four sides equal.

76. The number of lines of symmetry in a rectangle is:
   [A]   1   [B]   2   [C]   4   [D]   8

77. Symmetrically, a square differs from a rectangle because:
   [A]   A square has 2 lines of symmetry while a rectangle has 4.
   [B]   A square has 4 lines of symmetry while a rectangle has 2.
   [C]   The 4 sides of a square are equal but a rectangle has a pair of opposite
         sides equal.
   [D]   The diagonals of a square intersect at right angles but those of a rectangle
         do not.

78. The shapes in figure I to V, which are mirror images of each other, are:

      I.   II.   III.   IV.   V.
   [A]   I and II   [B]   III and V   [C]   I and IV   [D]   II and IV

79. The figure below which possesses rotational symmetry of order greater than one
    is:
   [A]   [B]

   [C]   [D]

80. The figure below which has no line symmetry is:

71

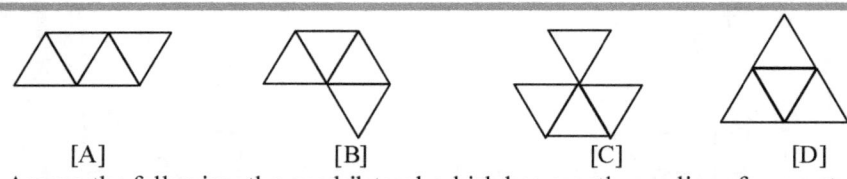

| [A] | [B] | [C] | [D] |

81. Among the following, the quadrilateral, which has exactly one line of symmetry, is:

   [A] A kite   [B] A rectangle   [C] A Parallelogram   [D]   A Rhombus

82. The number of lines of symmetry in the triangle below is:

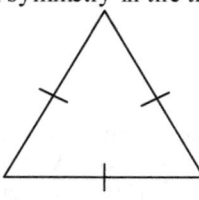

   [A] 2   [B] 3   [C] 4   [D] 5

83. The figure, which has two lines of symmetry, is:

   [A]   An isosceles triangle   [B] A square
   [C]   An equilateral triangle   [D] A rhombus

84. The figure, which has 9 planes of symmetry, is:

   [A] A regular nonagon   [B] A cube
   [C] A regular octagon   [D] A regular hexagon

85. The diagrams below which is not symmetrical about a horizontal axis is:

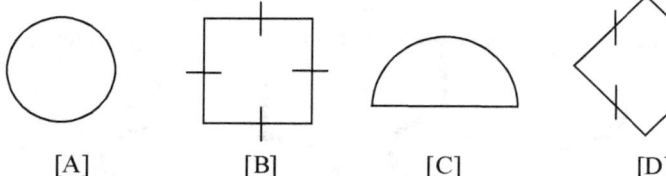

   [A]   [B]   [C]   [D]

86. A plane figure may have:

   [A] A line of symmetry and a plane of symmetry.
   [B] A line of symmetry and a point of symmetry.
   [C] A point of symmetry and a plane of symmetry.
   [D] An axis of symmetry and a plane of symmetry.

87. A solid figure may have:

   [A] A line of symmetry and a plane of symmetry.
   [B] A line of symmetry and a point of symmetry.
   [C] A point of symmetry and a plane of symmetry.
   [D] An axis of symmetry and a plane of symmetry.

Module 19

Statistics and Probability

Family of Situations
Module 19 is an extension of module 4, 8 and 13. At the end of the module, the student is expected to have acquired many more competencies within the **family of situations** *'Organization of Information and Estimation of Quantities in the Consumption of Goods and Services'*.

Categories of Action
The categories of action for module 19 include:
1. Organization, presentation and exploitation of information;
2. Interpretation of results.
3. Taking Chances

Credit
The module is expected to be covered within 3 weeks teaching 4 hours per week (or within 10 to 12 hours).

Topic 3

STATISTICS

Objectives

At the end of this topic, the learner should be able to:

1. Draw up frequency distribution tables.
2. Represent data using a bar chart, a pie chart or a histogram.
3. Read and interpret data from charts.
4. Find class width, mid class value or centre.
5. Find the measures of central tendencies for given data.
6. Construct histograms for group data.

We begin this module by revising the work done in form 1 module 4 and form 2 module 8. The student may rush to this section and revise it before continuing.

Review and Revision Exercise

A boy measured to the nearest metre how far he could through a tennis ball. The results are as follows.

66 69 70 68 71 68 69 70 67 68
67 68 67 66 69 68 69 70 68 67

Tally the raw data and represent the data on the following
(a) Frequency distribution table.   (b) Bar chart   (c) Pie char

3.1   Histograms

Histograms are similar to chronological bar charts. The only difference is that while the frequency is proportional to the length of the bars in the case of the bar chart, the frequency is proportional to the areas of the rectangles in the case of the histogram. Therefore, for the histogram, both the length and the width of the rectangle are important. Though the width of the rectangle, may be of different sizes it is often more convenient to make them the same size. In this way, histograms are therefore very similar to chronological bar charts

Example

The following table shows the number of form 2 students in a certain school in the year 2002 who had the required textbooks for the subjects Mathematics (M), English (E), French (F), History (H), Geography (G), Chemistry (C), Physics (P), Biology (B) and Literature (L). Draw a histogram to represent this information.

Textbook	M	E	F	H	G	C	P	B	L
No. of students	14	15	11	3	9	5	7	1	13

Solution

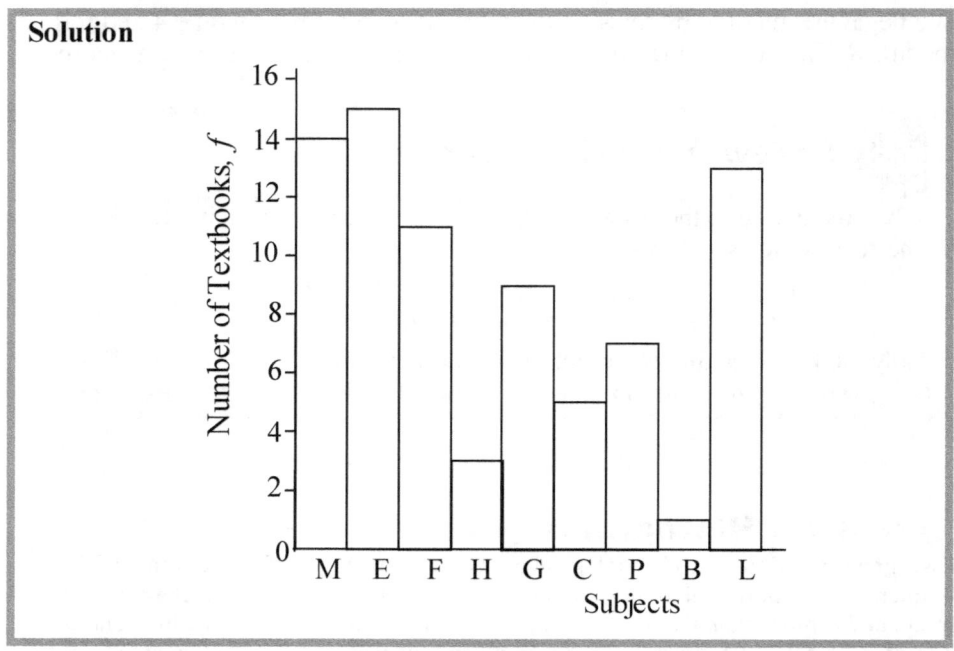

3.2 Measures of Central Tendencies

Mode, mean and **median**, which are very commonly used, are examples of averages, otherwise called measures of central tendencies. We call them measures of central tendencies because their values are representative or typical of any given data, and tend to lie centrally when we rank or arrange the data in order of magnitude (from highest to smallest or smallest to highest). Each of these measures has its advantages and disadvantages depending on the data and it intended purpose. For instance, the mean has the disadvantage that extreme values strongly affect it. Extreme values do not affect the median. On the other hand, the mode may or may not exist and turns to be very subjective. At times, there are many modes.

3.3 Mode

The mode of any given data is the variable or statistic that occurs most frequently.

 Example

1. Find the mode of the data. 1,2,4,6,2,7,7,2,2,7.

Solution

x	1	2	6	7
f	1	4	1	3

The frequency of 2 is 4, which is the highest frequency, so 2 is the mode.

2. Find the mode of the following data.
    (i) 99, 100, 101, 102 and 101. (ii) 99, 100, 101, 100, 102 and 101.

    Solution
    (i) Mode =101   (ii) Mode =100 and 101

Notice in the example above(ii) that there are 2 modes, 100 and 101. We call such a distribution a **bimodal** distribution. If there are three modes, we call the distribution a **trimodal** distribution and generally, if there are many modes we call the distribution a **multimodal** distribution.

3.4   Arithmetic Mean (Average or Mean)

We denote the mean by $\overline{x}$ and obtain it by summing all the data and dividing by the frequency. Thus

$$\overline{x} = \frac{\text{sum of data}}{\text{total frequency}}$$

$$\overline{x} = \frac{\sum x}{\sum f}$$

Where $\sum x$ and $\sum f$, read 'summation x' and 'summation f', respectively meaning 'sum of data' and 'sum of the frequencies' respectively. $\sum$ is called the sigma notation and $\sum x$ and $\sum f$, are read 'sigma x' and 'sigma f' respectively.

 Example

Find the mean of 11, 9, 15, 12 and 13

Solution

$$\overline{x} = \frac{\sum x}{\sum f} = \frac{11+9+15+12+13}{5} = 12$$

If some of the data repeat themselves, we take advantage of multiplication as repeated addition, to write the formula as

$$\bar{x} = \frac{\sum xf}{\sum f}$$

$\sum fx$, is read 'sigma fx', or 'summation fx', where fx means the product of each statistic and its frequency.

 Example

1. The following shows the marks obtained by 30 students during a test. Calculate the average mark.

55	60	65	40	60	60
65	50	40	60	50	60
60	50	60	30	40	60
60	50	60	50	60	50
60	50	60	60	50	60

Solution

To ease the work we draw a frequency distribution table.

Mark, x	Frequency, f	fx
30	1	30
40	3	120
50	8	400
55	1	55
60	15	900
65	2	130
	$\sum f = 30$	$\sum fx = 1635$

$$\bar{x} = \frac{\sum fx}{\sum f} = \frac{1635}{30} = 54.5$$

2. Find the mean of the following data.
    13, 13, 13, 13, 13, 13, 14, 14, 15, 15,
    15, 16, 16, 16, 16, 16, 16, 16, 16, 16

Solution

x	f	fx
13	6	78
14	2	28
15	3	45
16	9	144
	$\sum f = 20$	$\sum fx = 295$

$$\bar{x} = \frac{\sum fx}{\sum f} = \frac{295}{20} = 14.75$$

3.5   Median

To obtain the median, first rank the data. In other words, arrange the data in order of magnitude. For an odd number of numbers, the median is the middle number and for an even number of numbers, the median is the average of the two middle numbers.

Example

1.   Find is the median of 12,2,7,13,6.

   Solution
   Ranking: 2, 6, 7, 12, 13
   ∴Median = 7

2.   Find the median of 2, 7, 6, 13, 12, and 8

   Solution
   Ranking: 2, 6, 7, 8, 12, 13
   $$\therefore \text{ median} = \frac{7 + 8}{2} = 7.5$$

Exercise 3:1

1. Find the number that must be removed from the eight numbers 4, 11, 13, 8, 4, 5, 8 and 2, so that the mean of the remaining seven numbers is 6.
2. The mean of five numbers is 4. When we add a sixth number, the mean of the six numbers is $3\frac{1}{2}$. Find the sixth number.
3. Given that the mean of 3, 4 and m is 6, find the mean of 2, m and 14.
4. The following table represents the weights in kg of 11 students.

Weight, kg	45	53	54	49
No. of students	2	3	4	2

(a) State the modal weight of the students.
(b) Find the mean weight of the students.
(c) Find the median of the distribution.

5. Use the frequency distribution in the table below to calculate:
   (a) the mean   (b) the modal score
   (c) Find the median of the distribution.

Score (x)	1	2	3	4	5	6
Frequency (f)	4	6	7	3	3	1

6. The following table shows the marks obtained by pupils in a mathematics test.

Marks (x)	0	3	5	6	8	9	10
No. of pupils (f)	2	4	6	2	4	1	1

(a) State the mode of the distribution.
(b) Calculate, to 1 decimal place, the mean of the distribution.
(c) Find the median of the distribution.

7. Consider the frequency distribution below.

Score x	3	5	7	9	11
Frequency f	4	6	10	5	5

(a) State the mode of the distribution.
(b) Calculate, to 1 decimal place, the mean of the distribution.
(c) Find the median of the distribution.

8. The table below shows the number of coins of six denominations in a bag. Find:
   (a) the average value of the coins in the bag.
   (b) the mode of the coins in the bag.
   (c) the median of the distribution.

Value of coin FRS	5	10	25	50	100	500
Number	4	10	6	8	15	7

9. The weights of 8 teachers in a certain primary school were measured in kg as follows: 74,64,68,76,80,72,68 and 60 respectively. Find
   (a) the median.   (b) the mode of the data.

(c) their mean weight.

10. The frequency distribution in the table below shows the scores in a mathematics test in a certain class.

Score (x)	2	3	4	7	8	9
Frequency (f)	1	4	6	8	9	2

    (a) Find how many students wrote the test.
    (b) Find the mode of this distribution.
    (c) Find the mean mark for the test to 1 d.p.

11. The numbers of absences from a mathematics class registered within the first 20 lessons in the first term are 2, 3, 1, 0, 0, 4, 3, 2, 2, 2, 1, 4, 5, 5, 0, 0, 1, 1, 2, and 2. Find the
    (a) mode   (b) median   (c) mean number of absences.

12. 10 packets of different sizes contain sweets as shown in the table below.

Number of sweets	5	12	6	15
Number of packets	4	2	3	1

    (a) State the mode of the number of packets.
    (b) Find the median of the number of packets.
    (c) Calculate the mean number of sweets per packet.

3.6   Choicest Measure of Central Tendency

In choosing which measure of central tendency to use, we must consider two things. These are:

1. The nature of the distribution,
2. The purpose we intend to use the measure of central tendency.

Mean

The mean has the advantage that:
(a) We can express it using a simple formula.
(b) It takes account of all the values involved in the distribution.

The disadvantage of the mean is that it is highly affected by extreme values. Therefore, it is not a good measure of central tendency if the distribution involves one or more extreme values in one direction.
For instance in the data 3, 5, 8, 36, the mean is 13. This will be a very misleading measure of central tendency.

Median

The median has the advantages that:

(a) It is very easy to calculate.
(b) Extreme values do not affect the median because it is the middle of the distribution.

The median has the disadvantage that too many values in one direction greatly affect it.
In the data 1, 3, 4, 6, 17, 18, and 19, the median is 6. This value is too low because most of the values are small. Therefore, the median is not a good representation of the distribution in this case.

Mode

The mode has the advantage that:

(a) It is far easier to determine.
(b) Extreme values do not affect it.

The disadvantage of the mode is that, it is meaningless when there are several values having the highest frequency of occurrence. For instance
2, 5, 3, 2, 2, 7, 5, 8, 5, 7, 3, 9, 7, 3

Ranking and tabulating the data

x	2	3	5	7	8	9
f	3	3	3	3	1	1

The values 2, 3, 5, and 7 by definition are the modes, which has no significance.

 Example

---

1. What is the most appropriate measure of central tendency for the following distribution?
3, 4, 4, 4, 4, 4, 6, 7, 9, 13

   Solution
   The median is 4, the mode is 4, and the mean is 5.8.
   The mean is the best because it is almost central and therefore the best representation of the data.

3. The hourly wages of five employees in an office are 252 FRS, 396 FRS, 328 FRS, 920 FRS and 325 FRS. Find

(a) The median hourly wage.   (b) The mean hourly wage.
(c) Comment on your result.

Solution

(a) Arranging in ascending order, the wages are 252 FRS, 328 FRS, 375 FRS, 396 FRS and 920 FRS. This gives the median of 375 FRS.

(b) $\bar{x} = \dfrac{252+396+328+920+325}{5} = 444.2$

(c) The extreme value 920 FRS does not affect the median but affects the mean. In this case, the median is a better measure of the average hourly wage than the mean.

Exercise 3:4

1. Which measure of central tendency do you think the manager of New Life Supermarket will be most interested in? Give reasons for your answer.

2. In a certain week, a bus driver brought to his Patron the following balances: 20,000 FCFA, 20,000 FCFA, 22,000 FCFA, 24,000 FCFA, 24,000 FCFA, 36,000 FCFA, and 38,000 FCFA respectively on each day of the week. What measure of central tendency would you use to compare the balances, and why?

3. The data in the table below shows the number of students of LS1, who passed in the following subjects: Mathematics (M), Physics (P), Chemistry (C), Biology (B) and Further Mathematics (F).

Subject	M	P	C	B	F
No. of students	8	7	2	8	1

Which measure of central tendency would be the most appropriate for the analysis of this data? Give reasons for your answer.

4. The table below is a survey of the number of pigs owned by a group of farmers. One of Nga's two pigs just had a litter of 7 and 8 piglets respectively, so he has 16 pigs.

Number of piglets	0	1	2	4	16
Number of students	10	10	6	3	1

(a) Explain why the average should not be used as a measure of central tendency in this case.

(b) State the most appropriate measure of central tendency required in this case.

3.7 GROUPED DATA

To analyse very large masses of raw data, it is often necessary to distribute the data into **classes, class intervals** or **groups,** as shown in the following table, which is frequency distribution of the scores of 50 students in a test. We call this grouped data.

Mark, x	30-39	40-49	50-59
Frequency, f	10	14	26

Though data grouped in this way is easier to analyse, this method has a disadvantage in that the grouping destroys most of the original details of the information.

Class Limits and Class Boundaries

Consider the class 30-39. We call the smaller number, 30 the **lower class limit**, and the larger number 39, the **upper class limit.** Thus, 40 and 50 are the lower class limits, while 49 and 59, are the upper class limits for the classes 40-49 and 50-59 respectively.
If the processor rounded the data to the nearest whole number, the true class limits (called **class boundaries**) will actually be 29.5, 39.5, 49.5 and 59.5. For the class 30-39, the lower and upper class boundaries will be 29.5 and 39.5 respectively. For the class 40-49, the lower and upper class boundaries will be 39.5 and 49.5 respectively.

Class Size

This is the difference between the upper class boundary and the lower class boundary. Other names for class size are **class width, class length** or class interval denoted by c.

Mid-Interval Value

Other names for the mid-interval value are class mark or mid-point. The mid-interval value is the arithmetic mean of the upper and lower class limits. Thus for the class 40-49,

$$\text{Class mark} = \frac{40+49}{2} = 44.5$$

The class mark is the mark, which is representative of the given class.
Therefore, for clarity and mathematical analyses, we assume that all observations within a given class coincide with the class mark.

3.8   HISTOGRAMS FOR GROUPED DATA

When drawing histograms for grouped data, it is preferable to use the class boundaries and the class mark than the class limits.

(a) *Histograms with equal class widths*

These types of histograms are the most commonly used. In this case, the frequency is proportional to the height (length) of the rectangles.

 Example

The table below shows the marks obtained by 80 students in an examination. Draw a histogram to represent this data.

Marks, x	Frequency, f
1-10	3
11-20	5
21-30	5
31-40	9
41-50	11
51-60	15
61-70	14
71-80	8
81-90	6
91-100	4

Solution

Marks Obtained by 80 Students in an Examination

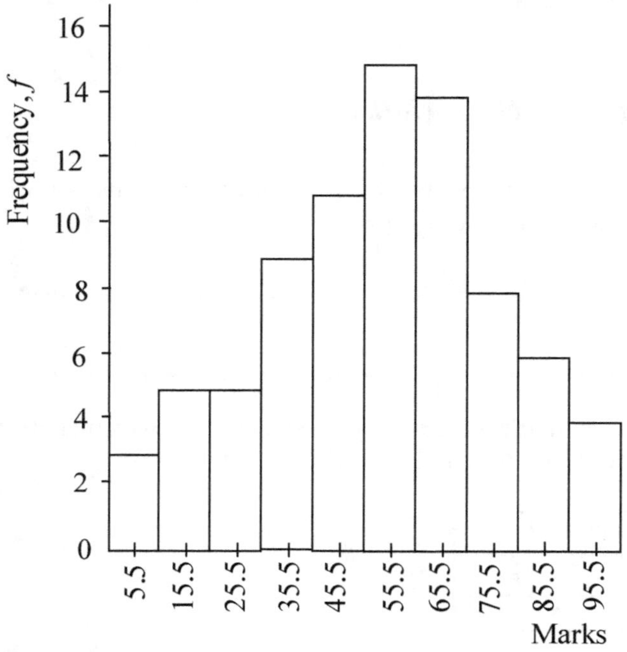

Finding the Mode from a Histogram

The Modal class is the class with the highest frequency represented by the tallest bar in the histogram.

Example

Use the histogram above to obtain the mode of the data above.

Solution
To obtain the mode we read the mark corresponding to the point of intersection of the dotted lines *AC* and *BD* as shown in the following figure. From the figure, the mode is 58.5.

Marks Obtained by 80 Students in an Examination

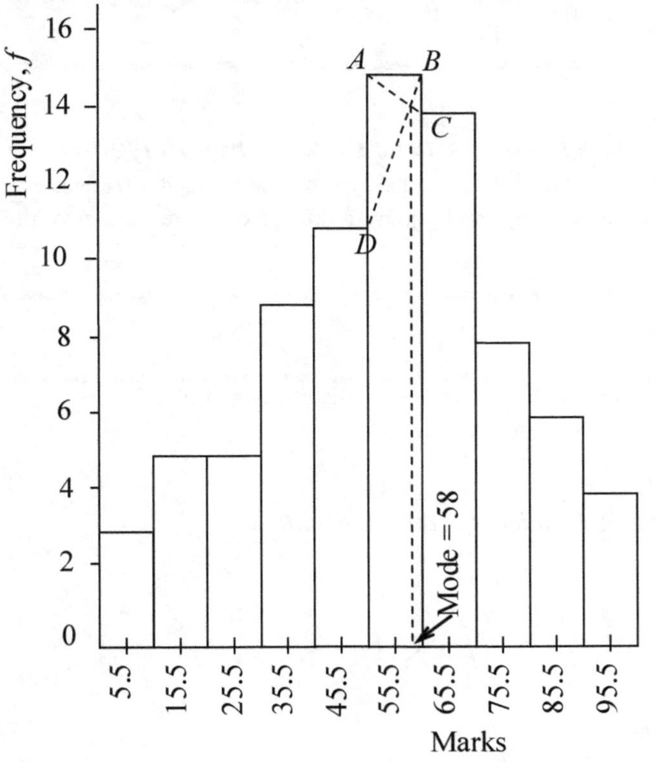

Finding the Mode by Calculation

If the class intervals are of equal sizes, we can obtain the mode using the formula

$$\text{mode} = L_1 + \left(\frac{\Delta_1}{\Delta_1 + \Delta_2} \right) c,$$

where

L_1 = lower class boundary of the modal class
C = class width
Δ_1 = modal class frequency − next lower class frequency
Δ_2 = next upper class frequency − modal class frequency

Example

By calculation, find the mode of the data above.

Solution

$$\text{mode} = L_1 + \left(\frac{\Delta_1}{\Delta_1 + \Delta_2}\right)c = 50.5 + \left(\frac{4}{4+1}\right)10 = 58.5$$

Frequency Distribution Curve (Frequency Polygon)

After drawing a histogram we can draw, another type of graph called a **frequency distribution curve** or **frequency polygon** by joining the tips of the rectangles of the histogram.

Example

1. Use the histogram above to draw a frequency distribution curve.

Solution

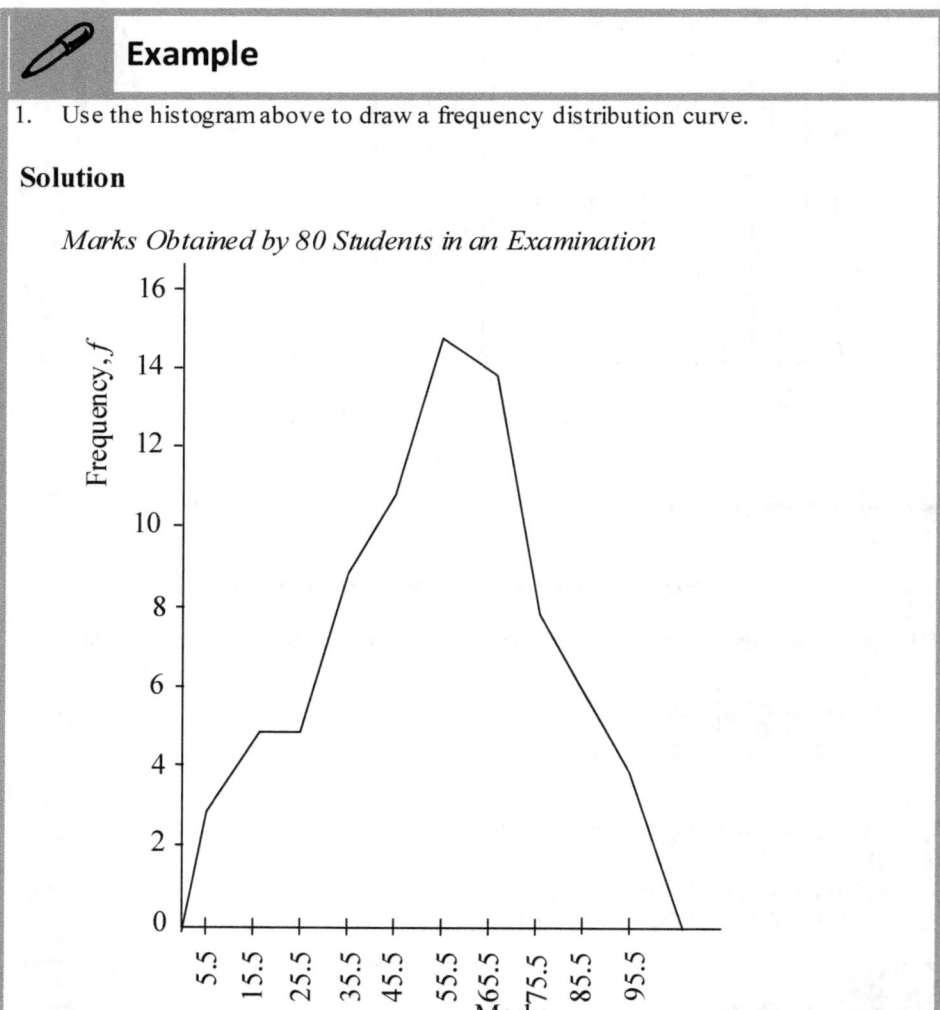

Marks Obtained by 80 Students in an Examination

2. Thirty-six students obtained the following scores out of 100 in a test.

25	49	76	12	51	56	81	50	45
92	58	67	55	52	43	31	48	84
66	56	44	39	45	22	56	74	98
67	34	41	34	68	69	70	85	51

(a) Starting with 0-9, arrange the marks in a grouped frequency table with class intervals of size 10.
(b) State the modal class.
(c) Draw a histogram to represent this data, hence obtain the mode of the distribution.
(d) Draw a frequency polygon of the distribution.

Solution

(a)

Marks, x	Class mark	Frequency, f
0-9	4.5	0
10-19	14.5	1
20-29	24.5	2
30-39	34.5	4
40-49	44.5	7
50-59	54.5	9
60-69	64.5	5
70-79	74.5	3
80-89	84.5	3
90-99	94.5	2

(b) The modal class is 50-59

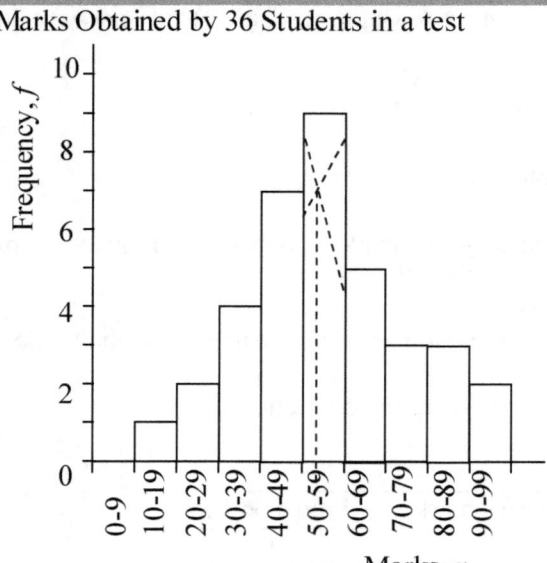

Marks Obtained by 36 Students in a test

(c) From the histogram, the mode is 53.5.
(d) The frequency polygon is as shown below.

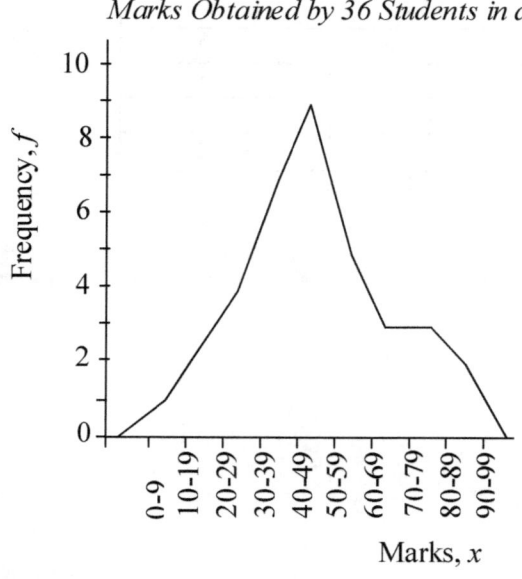

Marks Obtained by 36 Students in a test

(b) *Histograms with unequal class widths*

Histograms with unequal class widths are less common than histograms with equal class widths. However, it is important that we should take note of them. On such histograms,

Frequency = class width × standard frequency,

Or

$$\text{Standard frequency, S.F.} = \frac{\text{frequency of the class}}{\text{class width}}$$

Another name for standard frequency is relative frequency or frequency density and we can represent by the height of each rectangle. Therefore, the frequency is proportional to the area of the rectangle for each class and not to the height of each bar.

Example

The following table shows the wage distribution amongst three groups of employees in a large company.
(a) Draw a histogram for the distribution.
(b) Draw the frequency polygon for the distribution.

Wage in thousand CFA	Frequency
0-9	8
10-19	16
20-39	10

Solution

x	f	S.F.
0-9	8	0.8
10-19	16	1.6
20-39	10	0.5

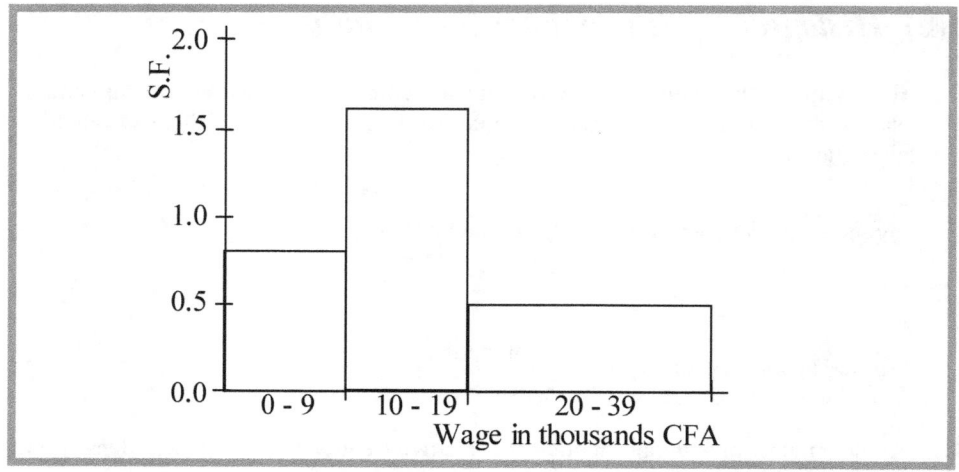

3.9 Cumulative Frequency

The cumulative frequency, C.F. is the total frequency of all the values less than the upper class boundary of a given class. For instance, in the distribution in the following example, the total frequency of all the classes up to and including the class 40-49 is $0 + 1 + 2 + 4 + 7 = 14$. This means that 14 students had marks less than 49.5. A **cumulative frequency table**, a **cumulative frequency distribution**, or simply a **cumulative distribution** is a table showing cumulative frequencies .

3.10 Cumulative Frequency Curves

Cumulative frequency curves, otherwise known as **ogives** are graphs of cumulative frequency against upper class boundary of each class.

Example

Draw a cumulative frequency table for the data in the table below.

Marks, x	Frequency, f
0-9	0
10-19	1
20-29	2
30-39	4
40-49	7
50-59	9
60-69	5
70-79	3
80-89	3
90-99	2

Solution

Marks, x	f	C.F.
< 9.5	0	0
< 19.5	1	1
< 29.5	2	3
< 39.5	4	7
< 49.5	7	14
< 59.5	9	23
< 69.5	5	28
< 79.5	3	31
< 89.5	3	34
< 99.5	2	36

3.11 Quantiles

Recall that given ranked data, the median M, is the middle term for an odd number of terms and the arithmetic mean of the two middle terms for an even number of terms. Thus, the median divides the data into two equal parts. We can extend this idea to those values, which divide the data into four, ten or even more equal parts.

We call the three values that divide data into 4 equal parts when we rank the data **quartiles**, denoted by Q_1, Q_2 and Q_3. The nine values, which divide data into 10 equal parts, when the data is ranked, are called **deciles**, denoted by D_1, D_2, D_3, D_4, D_5 ...D_9. In like manner, the ninety-nine values that divide data into 100 equal parts are called **percentiles** denoted by P_1, P_2, P_3, P_4...P_{99}.

The 50th percentile P_{50}, the 5th decile, D_5 and the 2nd quartile Q_2, all correspond to the median.

The 25th percentile P_{25} and the 75th percentile P_{75}, correspond to the first quartile Q_1 and the third quartile Q_3 respectively. Another name for the first quartile is the lower quartile and another name for the third quartile is the upper quartile. **Quantiles** is the general name for quartiles, deciles, percentiles and other values obtained by equal divisions of ranked data. We can easily obtain quantiles from cumulative frequency curves.

Example

The frequency distribution below shows the marks obtained by 80 students in an examination.

93

Marks, x	Frequency, f
1-10	3
11-20	4
21-30	6
31-40	8
41-50	12
51-60	15
61-70	13
71-80	9
81-90	6
91-100	4

(a) Make a cumulative frequency table for this distribution.
(b) Taking 1cm to represent 10 units on each axis, draw a graph of this distribution.
(c) Use your graph to obtain the
     (i) Median, M
     (ii) Lower quartile, Q_1
     (iii) Upper quartile, Q_3
     (iv) 90th percentile, P_{90}
     (v) 10th percentile, P_1

Solution

(a)

x	f	C.F.
≤ 10	3	3
≤ 20	4	7
≤ 30	6	13
≤ 40	8	21
≤ 50	12	33
≤ 60	15	48
≤ 70	13	61
≤ 80	9	70
≤ 90	6	76
≤ 100	4	80

(b) Marks out of 100 Obtained by 80 Students

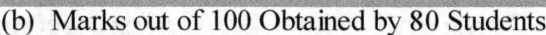

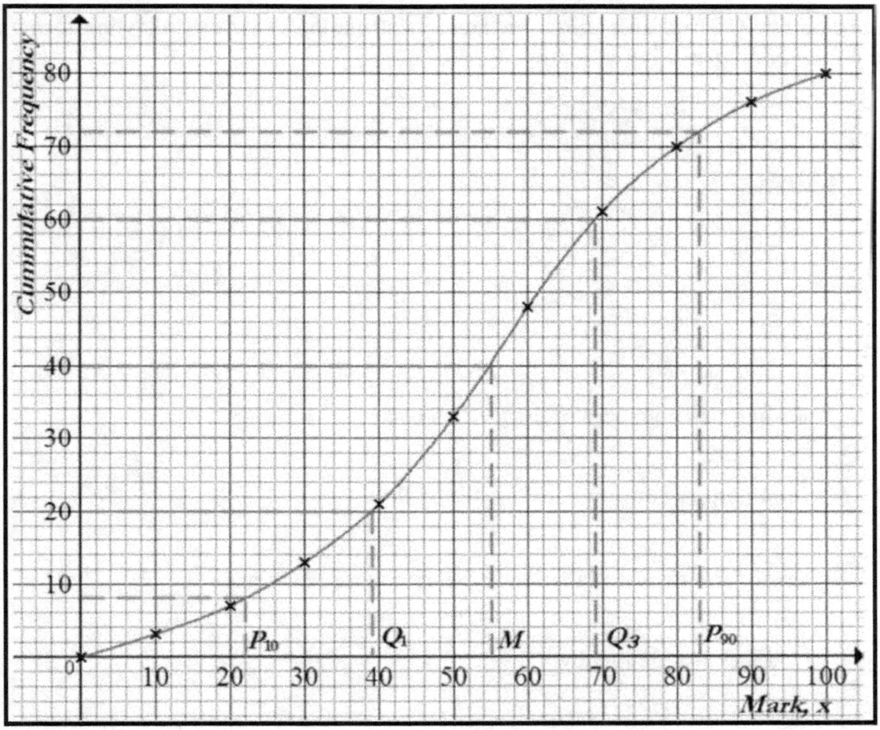

(c) From the graph above, we can read the median, the lower and upper quartiles, the 90th percentile and any other quantile from the x-axis by extrapolating from the y-axis. Thus;

(i) The cumulative frequency corresponding to median is $\frac{1}{2}$ of 80 = 40, therefore by extrapolation, the median M = 55

(ii) The cumulative frequency corresponding to the lower quartile is 25% of 80 = 20, therefore by extrapolation, the lower quartile, Q_1 = 39

(iii) The cumulative frequency corresponding to the upper quartile is 75% of 80 = 60, therefore by extrapolation, the upper quartile, Q_3 = 68

(iv) The cumulative frequency corresponding to 90th percentile is 90% of 80 = 72, therefore by extrapolation, the 90th percentile, P_{90} = 82.

(v) The cumulative frequency corresponding to 10th percentile is 10% of 80 = 8, therefore by extrapolation, the 10th percentile, P_{10} = 20

3.12 Median of Grouped Data by Calculation

We can also find the median of grouped data by calculation using the following formula, based on the principles of interpolation.

$$\text{Median} = L_1 + \left(\frac{\frac{\sum f}{2} - (\sum f)_1}{f_{median}} \right) c$$

Where,

L_1 = lower class boundary of the median class

$\sum f$ = total frequency

$(\sum f)_1$ = sum of frequency of all classes below the median class

f_{median} = median class frequency

c = median class size

 Example

Calculate the median of the data in the previous example.

Solution

$$\text{Median} = L_1 + \left(\frac{\frac{\sum f}{2} - (\sum f)_1}{f_{median}} \right) c$$

$$= 50.5 + \left(\frac{\frac{80}{2} - 33}{15} \right) 10 = 55.5$$

Measures of Dispersion (Variation, Spread or Scatter) of Data

The measures of dispersion are the degrees to which numerical data turns to spread about an average. The most common methods used to measure the spread of numerical data are the **range**, the **mean deviation**, the **inter quartile range** the **10-90 percentile range**, and the **standard deviation**.

The Range

The range of a set of data is the difference between the smallest and the largest statistic in the set.

 Example

Find the range of the following data. 2, 7, 3, 7, 8, 21, 17, 35, 4, 39.
Solution

Range = largest statistic − smallest statistic
= 39−2 = 37

Inter Quartile Range

The inter quartile range is defined as   Inter quartile range = $Q_3 - Q_1$

Semi-inter Quartile Range

We define the semi-inter quartile range as

$$\text{Semi-interquartile range} = \frac{Q_3 - Q_1}{2}$$

 Example

Use the graph above to find the inter quartile range and hence calculate the semi-inter quartile range of the data.

Solution

Inter quartile range = $Q_3 - Q_1$ = 68−38 = 30

Semi-interquartile range = $\dfrac{Q_3 - Q_1}{2} = \dfrac{30}{2} = 15$

The 10 - 90 percentile ranges

We define the 10-90 percentile range of a set of data as

$$\text{10-90 percentile range} = P_{90} - P_{10}$$

 Example

Use the cumulative frequency curve above to find the 10-90 percentile range of the data.

Solution

$$\text{10-90 percentile range} = P_{90} - P_{10} = 82 - 20 = 62$$

3.13 Mean of Grouped Data

We learnt earlier that the class mark is representative of any class. For any statistical analysis, we can assume all the observations within a given class to coincide with the class mark. To find the mean of grouped data therefore, we take the class mark to represent the mark for the whole class.

 Example

The marks x, obtained by 80 students in an examination are arranged as in the frequency table below. Determine the mean of the distribution.

Marks, x	Frequency, f
1-10	3
11-20	5
21-30	5
31-40	9
41-50	11
51-60	15
61-70	14
71-80	8
81-90	6
91-100	4

Solution

The class mark is chosen to represent the class.

Marks, x	Frequency, f	xf
5.5	3	16.5
15.5	5	77.5
25.5	5	127.5
35.5	9	319.5
45.5	11	500.5
55.5	15	832.5
65.5	14	917
75.5	8	604
85.5	6	513
95.5	4	382
	$\sum f = 80$	$\sum xf = 4290$

$$\bar{x} = \frac{\sum xf}{\sum f} = \frac{4290}{80} = 53.625 = 53.63 \text{ to 2d.ps.}$$

3.14 Mean Deviation from the Mean

As the name suggests, the mean (average) deviation from the mean is the mean of the deviations of all the data x_i from the mean. The mean deviation from the mean of a set of numbers $x_1, x_2, x_3, \cdots x_n$ denoted by *M.D.* is.

$$M.D. = \frac{\sum_{i=1}^{n} |x_i - \bar{x}|}{n}$$

Where $i = 1, 2, 3, \ldots n$ and $\bar{x}$ is the mean

$\sum_{i=1}^{n}$ Means the sum of the data from 1 to n

Example

Calculate the mean deviation from the mean of the set of numbers 2, 3, 6, 8, 11.

Solution

$$M.D. = \frac{\sum_{i=1}^{n}|x_i - \overline{x}|}{n}$$

$$\overline{x} = \frac{2+3+6+8+11}{5} = \frac{30}{5} = 6$$

$$\Rightarrow M.D. = \frac{|2-6| + |3-6| + |6-6| + |8-6| + |11-6|}{5}$$

$$= \frac{4+3+0+2+5}{5} = \frac{14}{5} = 2.8$$

If the frequency of any of the statistics in the data is greater than one, we can rewrite the formula above as

$$M.D. = \frac{\sum_{i=1}^{n} f_i|x_i - \overline{x}|}{\sum_{i-1}^{n} f_i}$$

Example

Find the mean deviation from the mean of the data in the previous example.
Solution
From the previous example the mean $\overline{x} = 53.625$.

Marks, x	f	$|x_i - \overline{x}|$	$f|x_i - \overline{x}|$
5.5	3	48.125	144.375
15.5	5	38.125	190.625
25.5	5	28.125	140.625
35.5	9	18.125	163.125
45.5	11	8.125	89.375
55.5	15	1.875	28.125
65.5	14	11.875	166.250
75.5	8	21.875	175.000
85.5	6	31.875	191.250
95.5	4	41.875	167.500
			1456.25

$$M.D. = \frac{\sum\limits_{i=1}^{5} f|x_i - \bar{x}|}{\sum\limits_{i=1}^{5} f} = \frac{1456.25}{80} = 18.20 \text{ to 2 d.p.s}$$

3.15 Standard Deviation and Variance

We denote the variance of a set of n numbers $x_1, x_2, x_3, \cdots, x_n$ with mean $\bar{x}$ by σ^2 or s^2 and define it as the mean of the squares of the deviations of each of the data from the mean. Thus,

$$\sigma^2 = \frac{\sum\limits_{i=1}^{n} (x_i - \bar{x})^2}{n}$$

We denote the standard deviation of a set of n numbers $x_1, x_2, x_3, ..., x_n$ with mean $\bar{x}$ by σ or s and define it as the square root of the variance (mean of the squares of the deviations of each of the data from the mean). Thus,

$$\sigma = \sqrt{\frac{\sum\limits_{i=1}^{n} (x_i - \bar{x})^2}{n}}$$

 Example

Find the variance and hence the standard deviation of the set of data 12, 6, 15, 7, 3, 10, 18, 5.

Solution

$$\sigma^2 = \frac{\sum\limits_{i=1}^{n} (x_i - \bar{x})^2}{n}, \quad \bar{x} = \frac{\sum\limits_{i=1}^{n} x_i}{n}.$$

x_i	$x_i - \bar{x}$	$(x_i - \bar{x})^2$
3	-6.5	42.25
5	-4.5	20.25
6	-3.5	12.25
7	-2.5	6.25
10	0.5	0.25
12	2.5	6.25
15	5.5	30.25
18	8.5	72.25
$\sum x_i = 76$		$\sum_{i=1}^{n}(x_i - \bar{x})^2 = 190$

$$\therefore \bar{x} = \frac{76}{8} = 9.5$$

We then use the mean $\bar{x}$ in the table to calculate $\bar{x} - x_i$ and $(\bar{x} - x_i)^2$.

$$\therefore \sigma^2 = \frac{190}{8} = 23.75 \text{ and } \sigma \approx 4.87$$

If the frequency of some of the statistics in the data is greater than one, we can rewrite the formula above as

$$\sigma = \sqrt{\frac{\sum_{i=1}^{n} f_i (x_i - \bar{x})^2}{\sum_{i=1}^{n} f_i}}$$

 Example

Find the variance and hence the standard deviation of the set of data 9, 3, 8, 8, 9, 8, 9, 18.

Solution

$$\sigma^2 = \frac{\displaystyle\sum_{i=1}^{n} f_i \left(x_i - \bar{x}\right)^2}{\displaystyle\sum_{i=1}^{n} f_i}, \quad \bar{x} = \frac{\displaystyle\sum_{i=1}^{n} f_i x_i}{\displaystyle\sum_{i=1}^{n} f_i}.$$

x_i	f_i	$x_i f_i$	$x_i - \bar{x}$	$f_i \left(x_i - \bar{x}\right)^2$
3	1	3	-6	36
8	3	24	-1	3
9	3	27	0	0
18	1	18	9	81
$\sum$	8	72		120

$$\therefore \bar{x} = \frac{72}{8} = 9$$

$$\therefore \sigma^2 = \frac{120}{8} = 15 \quad \text{and} \quad \sigma \approx 3.87 \text{ to 2 d.ps.}$$

3.16 SIMPLIFYING STATISTICS

At times, the data is so cumbersome that certain skills are required to simplify its analyses. These skills include:

1. Choosing a number called the *guess* or *assumed mean* or sometimes, *working zero*.
2. Using a method called *coding*.

1. Assumed or Guess Mean (Working Zero)

As the name may suggest, the assumed or guess mean is an arbitrary number, carefully chosen and assumed for the main time to be the mean. This number may be one of the statistics in the data or may not necessarily be.
The deviations d_i of the assumed mean A from any given statistic x_i in the data is defined as

$$d_i = x_i - A \Leftrightarrow x_i = A + d_i \Rightarrow \bar{x} = A + \bar{d}$$

103

By substituting $A + d_i$, for x_i we can transform all the statistical formulae so far met, the formulae into the easier usable equivalent forms below.

$$\text{Mean, } \bar{x} = A + \frac{\sum f_i d_i}{\sum f_i} = A + \bar{d}$$

$$M.D. = \frac{\sum f_i |d_i - \bar{d}|}{\sum f_i}$$

$$\text{Variance, } \sigma^2 = \frac{\sum f_i d_i^2}{\sum f_i} - \left(\frac{\sum f_i d_i}{\sum f_i}\right)^2$$

Where $\bar{d} = \dfrac{\sum f_i d_i}{\sum f_i}$. , is the mean of the deviations from the assumed mean.

Though these formulae at sight appear to be more complex than the previous ones, they are far easier to use if memorized.

 Example

In a class of 70 students, the following scores were recorded in a test.

Score	No. of students
4	3
5	6
6	5
7	6
8	7
9	10
10	13
11	8
12	7
13	2
14	3

Calculate
(a) the mean   (b) the mean deviation   (c) the standard deviation

Solution

$$\bar{x} = A + \frac{\sum f_i d_i}{\sum f_i} = A + \bar{d}, \quad M.D. = \frac{\sum f_i |d_i - \bar{d}|}{\sum f_i}$$

$$\sigma = \sqrt{\frac{\sum f_i d_i^2}{\sum f_i} - \left(\frac{\sum f_i d_i}{\sum f_i}\right)^2}$$

Let $A = 10$

The tabulation of the data is as in Table 37:36

(a) $\quad \bar{x} = 10 + \dfrac{-70}{70} = 9$

(b) $\quad \bar{d} = \dfrac{\sum fd}{\sum f} = \dfrac{-70}{70} = -1 \Rightarrow M.D. = \dfrac{146}{70} = 2.09$

(c) $\quad \sigma = \sqrt{\dfrac{532}{70} - (-1)^2} = \sqrt{7.6 - 1} = \sqrt{6.6} = 2.57$

x_i	f_i	d_i	$f_i d_i$	$|d_i - \bar{d}|$	$f_i |d_i - \bar{d}|$	d_i^2	$f_i d_i^2$
4	3	-6	-18	+5	15	36	108
5	6	-5	-30	+4	24	25	150
6	5	-4	-20	+3	15	16	80
7	6	-3	-18	+2	12	9	54
8	7	-2	-14	+1	7	4	28
9	10	-1	-10	0	0	1	10
10	13	0	0	1	13	0	0
11	8	1	8	2	16	1	8
12	7	2	14	3	21	4	28
13	2	3	6	4	8	9	18
14	3	4	12	5	15	16	48
$\sum$	70		-70		146		532

2. The Coding Method

To simplify the work more, we use the coding method for data grouped into a frequency distribution whose class intervals have equal size c. To use the coding method, we further transform the formulae into the following equivalent forms.

Mean, $\bar{x} = A + c\bar{u}$, where $\bar{u} = \dfrac{\sum fu}{\sum f}$

$M.D. = c\dfrac{\sum f|u - \bar{u}|}{\sum f}$

Standard deviation, $\sigma = c\sqrt{\dfrac{\sum fu^2}{\sum f} - \left(\dfrac{\sum fu}{\sum f}\right)^2}$

Example

Thirty-six students obtained the following scores out of 100 in a test.

Mark	Frequency
0-9	0
10-19	1
20-29	2
30-39	4
40-49	7
50-59	9
60-69	5
70-79	3
80-89	3
90-99	2

Calculate
(a) the mean
(b) the mean deviation
(c) the standard deviation to 2 d.p.s

Solution

$\bar{x} = A + c\bar{u}, \quad M.D. = c\dfrac{\sum f|u - \bar{u}|}{\sum f},$

$\sigma = c\sqrt{\dfrac{\sum fu^2}{\sum f} - \left(\bar{u}\right)^2}, \quad \text{where } \bar{u} = \dfrac{\sum fu}{\sum f}$

We build the following table so that we can obtain all the required sums in the formulae from it.

Mark	x	d	u	u^2	f	fu	fu^2	$|u-\bar{u}|$	$f|u-\bar{u}|$
0-9	4.5	−50	−5	25	0	0	0	5.08	0
10-19	14.5	−40	−4	16	1	−4	16	4.08	4.08
20-29	24.5	−30	−3	9	2	−6	18	3.08	6.17
30-39	34.5	−20	−2	4	4	−8	16	2.08	8.33
40-49	44.5	−10	−1	1	7	−7	7	1.08	7.58
50-59	54.5	0	0	0	9	0	0	0.08	0.75
60-69	64.5	10	1	1	5	5	5	0.92	4.58
70-79	74.5	20	2	4	3	6	12	1.92	5.75
80-89	84.5	30	3	9	3	9	27	2.92	8.75
90-99	94.5	40	4	16	2	8	32	3.92	7.83
Σ					36	3	133		53.82

(a) $\bar{x} = 54.5 + 10\left(\dfrac{3}{36}\right) = 55.33$ to 2d.p.s

(b) $\bar{u} = \dfrac{\sum fu}{\sum f} = 0.0833$

$\Rightarrow$ M. D. $= \dfrac{10(53.82)}{36} = 14.95$ to 2 d.p.s

(c) $\therefore \sigma = 10\sqrt{\dfrac{133}{36} - \left(\dfrac{3}{36}\right)^2} = 10\sqrt{3.6944 - 0.0069}$

$= 10\sqrt{3.686} = 10(1.920) = 19.20$ to 2 d.p.s

 ## Exercise 3:5

1. A boy measured to the nearest metre, how far he could throw tennis on 20 successive trials, and obtained the following results:

    66 69 70 68 71 68 69 70 67 68

    68 68 67 66 69 68 69 70 68 67

    (a) Draw a frequency distribution table for the data.
    (b) State the mode of the distribution.
    (c) Find the median of the distribution
    (d) An approximate formula for determining the mean is

    Mean − Mode = 3(Mean − Median)

    Use this formula to calculate the mean of the distribution.

    (e) Calculate the exact mean of the distribution, to one decimal place.

2. $\dfrac{1}{9}$ of the candidates obtained grade D and an equal number of the

candidates were awarded grades E and F.

(a) Copy and complete the following table that is used to draw up a pie chart.

GRADE	A	B	C	D	E	F
Angle of Sector					70	70

(b) Draw the pie chart accurately on a circle of radius 5 cm, labeling the sectors.

(c) Given that 6 candidates obtained grade A,
  (i) Find the number of candidates that took the test.
  (ii) Find the number that obtained grade D.

(d) Given that grades A, B, and C, are pass grades, state the ratio 'pass to fail' in its simplest form.

(e) The figure below shows the distribution of the grades for boys. On a similar set of axes, construct a bar chart to show the distribution of the grades for girls.

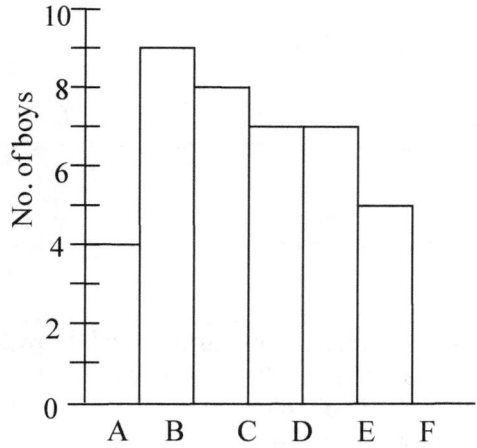

3. (i) The table below shows the group frequency distribution of examination marks for 120 students. Each mark is a whole number.
  (a) Construct a cumulative frequency table for this frequency distribution. Take the first class to be ≤10.

Marks	Number of candidates
1-10	0
11-20	2
21-30	6
31-40	7
41-50	14
51-60	20
61-70	35
71-80	29
81-90	6
91-100	1

(ii) Candidates with a score of 50 or less will have to re-sit the examination and those with scores above 60 will earn credit certificates.

    (b) Using your table or otherwise, determine, the number of candidates who

        (i)   Will have to re-sit for the examination.

        (ii)   Earned credit certificates.

4. A mathematics teacher gave a test and promised to award prizes to the top 30 students of the class. The marks scored by the top 30 students were recorded as follows:

10	8	4	4	5	7	4	6	5	4
9	7	6	4	8	8	6	7	4	6
8	5	6	4	8	7	6	4	6	4

    (a) draw a frequency table to show the distribution of marks over the top 30 students in the class

    (b) state the mode of the distribution

    (c) Calculate the mean mark for the top 30 students.

5. The points obtained by 40 players in a world soccer were recorded as follows:

82 42 61 57 55 39 67 78 65 66

22 71 67 8 45 49 56 52 68 14

60 57 18 64 58 38 50 46 83 76

44 29 61 34 74 81 91 48 47 59

  (a) Using the intervals 0-9, 10-19, 20-29, etc, construct a frequency table to show this information.

(b) State the modal class

(c) Making use of the mid-interval values calculate to 1 decimal place the mean of the distribution.

(d) Draw a cumulative frequency curve for this distribution.

(e) From your graph, find the inter-quartile range.

The organizers assigned the following grades to the above points.

≥ 70 (A grade), 50-69 (B grade),

30-49 (C grade), ≤ 30 (D grade).

(f) Eto'o Fils had 79 points. What was his grade?

6. The following numbers were marks gained by 50 pupils in an examination:

57	60	37	74	62	40	56	59	80	60
62	94	78	73	56	68	67	79	83	87
90	93	58	46	77	63	66	66	56	71
51	77	53	69	70	69	70	70	47	54
49	54	68	35	64	67	76	73	68	61

(a) Reduce these marks to a frequency distribution with equal intervals, having as the first interval 35-44 inclusive.

(b) Draw a cumulative frequency curve for this distribution.

(c) What is the median mark?

(d) Use your graph to estimate what percentage of candidates passed the examination if the pass mark is 55.

7. The mean of two numbers in a set A is 21.8. The mean of three numbers in a set B is 28.8. Find:

(a) the sum of the two numbers in set A,

(b) the sum of the three numbers in set B,

(c) the mean of the five numbers put together.

(d) When we introduce a sixth number x into the set in (c) above, the mean of the six numbers is now 20, find x.

8. The table below shows the time a sales girl used in serving 80 customers in a certain week.

Time (min)	Number of customers
$20 < t = 25$	8
$25 < t = 30$	8
$30 < t = 35$	12
$35 < t = 40$	30
$40 < t = 45$	18
$45 < t = 5\,0$	4

(i)   Calculate the mean time she spent for each customer.

(ii) Draw a cumulative frequency table and use it to draw the cumulative frequency curve for the above data, using a scale of 1 cm to represent 5 minutes on the time axis and 1 cm to represent 10 customers.

(iii) From your graph, determine the maximum time used to serve 75% of the workers.

9. The table below shows the time in minutes taken by 100 students for a quiz.

Time	11-20	21-30	31-40	41-50	51-60	61-70
Freq.	6	14	22	36	20	2

Find to 3 significant figures, the mean and standard deviation of the distribution.

10. The table below in a recruitment examination, the marks obtained out of 100 by 70 students. Calculate to two decimal places

(a) the mean of the distribution.

(b) the standard deviation of the distribution.

Mark	Frequency
1-10	3
11-20	5
21-30	9
31-40	14
41-50	12
51-60	9
61-70	7
71-80	6
81-90	5
91-100	0

11. The following table shows the number of times the 200 students in a certain school worked punishment. Arrange this data in class intervals 0-4, 5-9, 10-14 etc. then calculate

(a) the mean   (b) variance   (c) standard deviation.

No. of times	No. of students
0	34
6	1
10	3
12	4
15	6
16	4
17	3
18	5
19	2
20	11
21	3
22	6
23	8
24	13
25	20
26	5
27	6
28	9
29	5
30	12
31	4
32	5
33	2
34	3
35	10
36	8
37	2
38	1
40	4
44	1

12. The following table shows the diameters in mm of 50 watermelons.

Diameter	Number of fruits
160-161	1
162-163	3
164-165	9
166-167	20
168-169	14
170-171	2
172-173	1
174-175	0

Calculate the mean, median and standard deviation of the diameter.

13. The following table shows the time taken by 60 competitors to complete a given task. Estimate to one decimal place
    (a) the mean   (b) variance
    (c) standard deviation of the distribution.

Time	Frequency
10-19	3
20-29	7
30-39	12
40-49	18
50-59	10
60-69	6
70-79	3
80-89	1

14. The number of civil servants in a certain community classified by age groups is as in the table below.

Age	Frequency
15-19	66
20-24	65
25-29	56
30-34	50
35-39	42
40-44	37
45-49	35
50-54	30
55-59	24
60-64	22

Calculate;
(a) the arithmetic mean   (b) the median
(c) the mode   (d) the variance
(e) the standard deviation of the ages of the civil servants.

15. The following table shows the marks obtained by 500 candidates in an examination.

Mark	Frequency
20-29	66
30-39	65
40-49	56
50-59	50
60-69	42
70-79	37
80-89	35

(a) Calculate the mean mark of the distribution.

(b) Make a table of cumulative frequencies for marks below 29.5, 39.5 etc.

(c) Draw a cumulative frequency graph and use it to find;

(i)   the median mark

(ii)   the approximate number of candidates who attained a mark of at least 55 %.

 ## Multiple Choice Exercise 3

1. We call a survey of people and/or property:
    [A] a census   [B] data   [C] a population   [D] a sample

2. The tally marks ⊮⊮ ⊮⊮ ⊮⊮ ⊮⊮ ⊮⊮ ||| represent the number:
    [A] 18 [B] 23   [C] 28   [D] 33

3. The correct tally representation of 17 students is:
    [A] ⊮⊮ ⊮⊮ ⊮⊮   [B] ⊮⊮ ⊮⊮ ||||||| [C] ||||| |||| |||| || [D]
    ⊮⊮ ⊮⊮ ⊮⊮ ||

4. The tally marks ⊮⊮ ⊮⊮ ⊮⊮ ||| stand for:
    [A] 18 [B] 15   [C] 13   [D] 20

5. A school awarded 480 out of 720 candidates a D Grade. On a pie chart showing all the grades the angle at the centre for the D grade is:
    [A] 270° [B] 240° [C] 210° [D] 180°

6. The measure of central tendency a shoe company will be most interested in is:
    [A] mean   [B] mode   [C] median   [D] mean and median

7. The average of 0, 1, 6, 7, 9 and 19 is:
    [A] 9   [B] 6   [C] 7   [D] 10

8. The average of 1, 2, 5, 7, and 15 is:
    [A] 6   [B] 30   [C] 7   [D] 15

114

9. A group of four people found that their heights were 1.38 m, 1.71 m, 1.23 m and 1.40 m. Their average height (in metres) is:
   [A] 1.145   [B] 1.18   [C] 1.39   [D] 1.43

10. The average wage bill in FCFA of 40 men who collectively earn 3,540,000 FCFA is:
    [A] 87,000   [B] 29,500   [C] 88,500   [D] 31,700

11. The mean of 9,13,16,17,19,23,24 correct to two decimal places is:
    [A] 23.00   [B] 17.29   [C] 16.50   [D] 16.33

12. The average of the first four prime numbers greater than 10 is:
    [A] 20   [B] 19   [C] 17   [D] 15

13. The mean of 20 observations is 4. The observed largest value 23 is removed. The mean of the remaining observations is:
    [A] 4   [B] 3   [C] 2.85   [D] 2.60

14. The mean heights of the three groups of students consisting respectively of 20, 16 and 14 students are 1.67 m, 1.50 m and 1.40 m respectively. The mean height of all the students is:
    [A] 1.52 m   [B] 1.53 m   [C] 1.54 m   [D] 1.55 m

15. The mean of 30 observations is 5. The observed largest value of 34 is deleted. The mean of the remaining observations is:
    [A] 4   [B] 3.8   [C] 3.4   [D] 5

16. The following table shows the scores of some students in a test. The average score is 3.5. The value of x is:

Scores	1	2	3	4	5	6
No. of students	1	4	5	6	x	2

   [A] 1   [B] 2   [C] 3   [D] 4

17. The mean of 9, x and 13 is 11. The value of x is:
    [A] 7   [B] 8   [C] 11   [D] 13

18. The value of x which qualifies 4 as the mean of the data 4, $3x$, 0 and 3 is:
    [A] 1   [B] 2   [C] 3   [D] 4

19. The mean of five observations is 15. Four of them are 11, 12, 19 and 20. The fifth is:
    [A] 10   [B] 25   [C] 20   [D] 13

20. A pie chart is drawn to represent the percentages 20%, 50%, 25% and 5%. The angle which represents 5% is:
    [A] 5°   [B] 18°   [C] 25°   [D] 126°

21. Given the scores −3, 4, 0, 4,−2,−5, 1, 7,10,5 the median of the scores is:
    [A] 2.5   [B] 2   [C] 4   [D] 3.5

22. From the table below, the mean number of male children per family is:
    [A] 5   [B] 4   [C] 3   [D] 2

115

No. of male children	0	1	2	3	4
No. of families		4	8	6	2	7

23. The mean of four numbers a, b, c and d is 6. The mean of 5 numbers a, b, c, d and e is 10. The value of e is:
    [A]   24   [B]   25   [C]   26   [D]   27

24. The average age of five boys is 11 years. A sixth boy whose age is 17 years is added, the mean age in years will now be:
    [A] 14   [B] 12   [C] 13   [D] 11

25. The median of 8, 10, 9, 6, 7, 10, 12, 8, 9, 8 is:
    [A]   7.5   [B]   8   [C]   8.5   [D]   8.7

26. The median of the set of scores 65, 75, 55, 48, 78 is:
    [A]   55   [B]   60   [C]   72   [D]   65

27. The median of the set of numbers 2.64, 2.50, 2.72, 2.91, 2.35 is:
    [A]   2.72   [B]   2.64   [C]   2.50   [D]   2.35

28. Given the set of numbers 12,15,13,14,12 and 12. The median is:
    [A]   12.5   [B]   12   [C]   13   [D]   13.5

29. The following table shows the age distribution of a group of children. Their median age is:
    [A]   4 years   [B] 7 years   [C] 8 years   [D] 9 years

Age (in years)	4	5	6	7	8	9	10
Frequency	2	1	2	4	3	6	2

30. The following table gives the marks scored by a group of students in a test. The median mark is:
    [A] 4   [B]   3   [C]   2   [D]   1

Mark	0	1	2	3	4	5
Frequency	1	2	7	5	4	3

31. The mode of the numbers 8, 10, 9, 9, 10, 8, 11, 8, 10, 9, 8 and 14 is:
    [A] 8   [B]   9   [C]   10   [D]   11

32. A group of students measured a certain angle (to the nearest degree) and obtained the following results.
    $75°$ $76°$ $72°$ $73°$ $74°$ $79°$ $72°$
    $72°$ $77°$ $72°$ $71°$ $70°$ $78°$ $73°$

    The mode of their measurements is:
    [A] $78°$ [B] $74°$ [C] $73°$ [D] $72°$

33. The measure, which is not a measure of dispersion, is:
    [A] Mode   [B]   mean deviation
    [C] Inter-quartile range   [D]   standard deviation

34. It is true to say that the measure, which is not measure of dispersion, is:

[A] Range   [B] Variance   [C] Mode   [D] Percentile range

35. The Variance of a given distribution is 25. The standard deviation is:
    [A] 625 [B] 75 [C] 25 [D] 5

36. The standard deviation of the marks 2, 3, 6, 2, 5, 0, 4, 2 is:
    [A] 1.5 [B] 1.7 [C] 1.8 [D] 1.9

37. The standard deviation of the numbers 2, 5, 6, 4 and 8 is:
    [A] 2   [B] 4   [C] 6   [D] 7

38. The mode of the distribution in the table below is:
    [A] 2   [B] 3   [C] 4   [D] 5

Score	0	1	2	3	4	5
Frequency	2	3	4	2	7	2

39. The mean score of the distribution in the table above is:
    [A] 1.75 [B] 2 [C] 2.5 [D] 2.75

40. The median score of the distribution in table above is:
    [A] 0   [B] 2.5   [C] 3   [D] 5

41. For a class of 30 students, the scores in a test out of 10 marks were as in the table below. The mode is:
    [A] 3   [B] 5   [C] 6   [D] 7

4	5	7	2	3	6	5	5	8	9
5	4	2	3	7	9	8	7	7	7
3	4	5	5	2	3	6	7	7	2

42. For a class of 30 students, the scores in a test out of 10 marks were as in the table above. The median score is:
    [A] 3   [B] 5   [C] 6   [D] 7

43. For a class of 30 students, the scores in a test out of 10 marks were as in the table above. The range of the distribution is:
    [A] 7   [B] 2   [C] 8   [D] 9

44. The table below, shows the tithes in thousand FCFA, collected in a church. The mode is:
    [A] 3   [B] 6   [C] 9   [D] 12

Amount (thousand FCFA)	3	6	9	12	15	18
No. of Christians	3	9	6	15	3	12

45. The table above shows the tithes in thousand FCFA, collected in a church. The median of the distribution is:
    [A] 3   [B] 9   [C] 12   [D] 15

46. The table above shows the frequency distribution of a number of chairs in each rooms of a hotel. The mean of the distribution is:
    [A] 3.5 [B] 4.0 [C] 4.4 [D] 5.0

47. The table above shows the frequency distribution of a number of chairs

in each rooms of a hotel. The mode of the distribution is:

[A] 2   [B] 5   [C] 7   [D] 9

48. The table above shows the frequency distribution of a number of chairs in each rooms of a hotel. The median of the distribution is:

[A] 4   [B] 4.5   [C] 5   [D] 5.5

49. The table below shows the frequency distribution of marks scored by a group of students in a test. The number of students who took the test is:

[A] 14   [B] 15   [C] 18   [D] 20

Marks	2	3	4	5	6
Frequency	2	4	5	3	1

50. The table above shows the frequency distribution of marks scored by a group of students in a test. The modal score is:

[A] 2   [B] 3   [C] 4   [D] 5

51. The table above shows the frequency distribution of marks scored by a group of students in a test. The mean mark is:

[A] 1.3   [B] 2   [C] 3   [D] 3.8

52. The table below shows the scores of 15 students in a physics test. The number of students who scored at least 5 is:

[A] 6   [B] 8   [C] 9   [D] 7

Marks	1	2	3	4	5	6	7	8	9	10
No. of students	1	3	2	0	1	6	1	0	1	0

53. The table above shows the scores of 15 students in a Physics test. The median score is:

[A] 5   [B] 6   [C] 7   [D] 8

54. The table below shows the scores of a group of 40 students in a Biology test. If the mode is m and the median is n then (m, n) as an ordered pair is:

[A] (5,5)   [B] (5,6)   [C] (6,5)   [D] (9,4)

Score	1	2	3	4	5	6	7	8	9
Frequency	2	3	6	7	9	6	2	2	3

55. The table above shows the scores of a group of 40 students in a physics test. The mean of the distribution is:

[A] 4.5   [B] 4.8   [C] 5.0   [D] 5.2

56. The table below shows the number of goals scored by a football team in 20 matches. The mean number of goals scored is:

[A] 1.75   [B] 1.9   [C] 2   [D] 2.15

Number of goals	0	1	2	3	4	5
Number of matches	3	5	7	4	1	0

57. The number of goals scored by a football team in 20 matches is shown in the table above. The modal number of goals scored is:
[A] 1   [B] 2   [C] 5   [D] 7

58. The distribution by Region of 840 students in the faculty of science of the University of Buea in a certain session is as follows:

Adamawa Region   45
North West Region   410
Littoral Region   105
Western Region   126
South West Region   154

In a pie chart drawn to represent this distribution, the angle subtended by Western Region is:
[A] 42°   [B] 45°   [C] 48°   [D] 54°

59. The pie chart below illustrates the amount of private time a student spends in a week studying various subjects. The value of k is:
[A] 15°   [B] 30°   [C] 60°   [D] 90°

60. The pie chart below illustrates the amount of private time a student spends in a week studying various subjects. Given that, he spends 2 and a half hours on science, the total number of hours he studies in a week is:
[A] $3\frac{1}{2}$ [B] 5   [C] 8   [D] 12

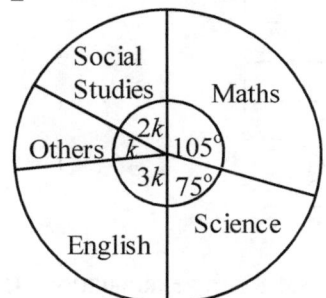

61. The marks obtained by pupils of a certain class are grouped as shown below; 0-4, 5-9, 10-14, 15-19. It is true to say that:
I: The mid values of the grouped marks are 2,7,12, and 17.
II: The class interval is 4.
III: The class boundaries are 0.5, 4.5, 9.5, 14.5 and 19.5.
[A] I only   [B] II only   [C] III only   [D] I and II

62. The pie chart below represent the number of fruits on display in a grocery shop. Given that there are 60 oranges in display, the number of apples is:

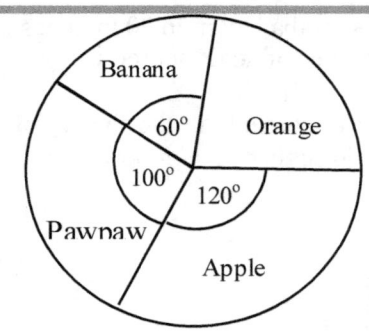

[A] 40   [B] 80   [C] 90   [D] 120

63. The histogram below shows the number of candidates, in thousands who obtained given ranges of marks in an entrance examination.

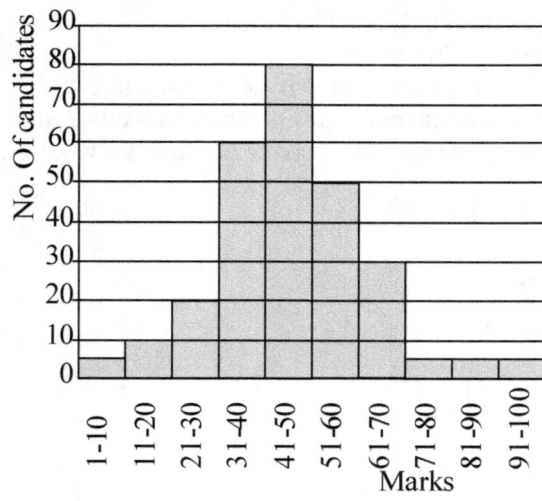

The total number of candidates who sat for the examination is:

[A] 120,000   [B] 250,000   [C] 260,000   [D] 270,000

64. The histogram above shows the number of candidates, in thousands who obtained given ranges of marks in an entrance examination. The number of candidates who scored at most 30% is:

[A] 20,000   [B] 25,000   [C] 35,000   [D] 60,000

65. The histogram below shows the distribution of a group of students according to their ages. The range of their ages is:

[A] 14 years   [B] 20 years   [C] 30.5 years   [D] 31 years

66. The histogram above shows the distribution of a group of students according to their ages. The mode of their ages is:

[A] 22.5 years   [B] 23.0 years   [C] 24.0 years   [D] 24.5 years

67. For six sequences, Ngange scored 76, 57, 97, 86, 86, 70 in Mathematics. If Ngange wants to convince his parents of his strength in Mathematics,

the measure he should use should be:
[A] Mean [B] median [C] mode [D] range

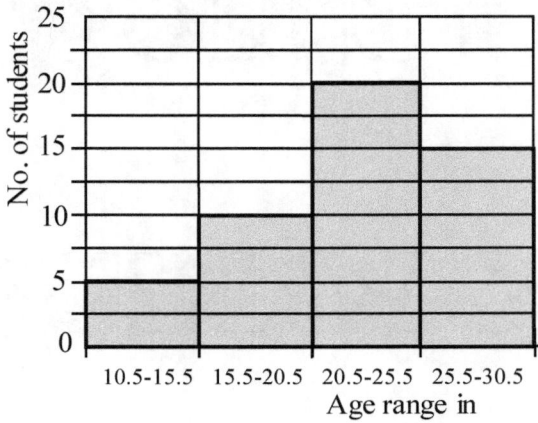

68. Six employees earn 800 FCFA, 850 FCFA, 900 FCFA, 950 FCFA, 1000 FCFA, and 2350 per hour. The manager claims that the median of the hourly wages is 925 FCFA. The manager is:
    [A] wrong because 925 FCFA is the mode.
    [B] wrong because he seems to ignore the amount 2350 FCFA.
    [C] wrong because 925 FCFA is the mean.
    [D] right because 925 FCFA is the correct median.

69. The president of a certain Credit Union used the data in the table below to find the mean monthly salary of the Credit Union staff.

Monthly Salary	No. of workers
26,000 FCFA	7
30,000 FCFA	8
240,000 FCFA	1
260,000 FCFA	1
300,000 FCFA	3

In a report the president said that, the typical salary at the Credit Union is about 92,000 FCFA. His statement is:
[A] misleading because salaries of five staff are far above those of the other fifteen.
[B] misleading because the mean of the data is not 92,000 FCFA.
[C] misleading because the president ignored the highest salary.
[D] Correct.

Topic 4

PROBABILITY

Objectives

At the end of this topic, the learner should be able to:

1. Define probability or say what probability is about.
2. Understand and the basic probability terminologies such as biased, unbiased, trial, event, outcome, sample space, event subset etc.
3. Calculate simple probabilities.
4. State and use the standard definition of probability.
5. Appreciate that probability is a number in the range $0 \leq P(E) \leq 1$ and that the probability of a sure event is 1, while that of an impossible event is 0.
6. Perform simple experiments using a die and/or playing cards.

4.1 Some Basic Probability Terminology

 Review and Revision Exercise

Match the following terms with the correct description in numbers 1 to 7.
Sample space, unbiased, biased, trial, event subset, event, outcome.
1. A coin which is as likely to turn up heads as to turn up tails is said to be.
2. If the likelihood of a coin turning up heads is not equal to the likelihood of turning up tails we say the coin is.
3. The name given to the act of tossing a coin or a die.
4. The term used to describe the appearance of a head or tail in the course of tossing a coin is.
5. One of the possible occurrences (*H* or *T* in the case of a coin), called.
6. The term used to describe the set of all the possible outcomes in a particular experiment is.
7. In an experiment, a teacher chooses a letter at random from the letters of the word 'MATHEMATICS'. What is the possibility space for the event of choosing a vowel?

4.2 Probability as a Number

The probabilities that any event *E* will occur (or will not occur) always lie between zero and one *inclusively*.
This means that the probability of an event occurring is always zero, one or any number between zero and 1. The probability of an event *E* occurring is denoted by $P(E)$.

Thus; $0 \le P(E) \le 1$

The probability of a **sure** or **certain** event *A* is 1.

$P(E) = 1 \Leftrightarrow E$ must occur.

For instance, the probability that any living person will die one day is 1. The probability of an event *B* occurring is zero, when it is **impossible** for the event to occur.

$P(E) = 0 \Leftrightarrow E$ cannot occur.

For instance, the probability that someday a man will be pregnant is zero.
The probability of an event *E* occurring is any number between 0 and 1 exclusively (not including 0 and 1). This means that there are some chances of the event occurring and some chance of the event not occurring. For instance if

it is stated that the probability of an event E occurring is $\dfrac{1}{4}$ it means that out of 4 trials, the event is expected to occur once and it is expected not to happen 3 times and out of 40 trials, the event is expected to happen 10 times and it is expected not to happen 30 times.

4.3   Equiprobable Outcomes

Equiprobable or **equally likely** events are events, which have equal chances of occurrence. If there are n such events, the probability $P(E)$ of one of the events occurring is given by $P(E) = \dfrac{1}{n}$.

 Example

1.   A fair coin is tossed once. State the probability of:
     (a) a head   (b) a tail

     Solution
     $S = \{H, T\}$

     (a) $P(H) = \dfrac{1}{2}$ (b) $P(T) = \dfrac{1}{2}$

     $\Rightarrow P(H) = P(T) = \dfrac{1}{2}$ or 0.5 or 50%

2.   State the probability of each of the faces showing 1, 2, 3, 4, 5, and 6 if a fair die is tossed.

     Solution
     $S = \{1, 2, 3, 4, 5, 6\}$

     $P(1) = P(2) = P(3) = P(4) = P(5) = P(6) = \dfrac{1}{6}$

4.4   Standard Definition of Probability

Suppose a sample space S consists of a finite number of equiprobable out comes. Then we define the probability of an event E occurring as

$$\text{Probability of } E = \frac{\text{No of outcomes in the event } E}{\text{Total number of outcomes } S}$$

i.e. $P(E) = \dfrac{n(E)}{n(S)}$

4.5　　Suits of Playing Cards

An ordinary pack of playing cards contains 52 cards. There are four types of cards; hearts, clubs, diamonds and spades; each type having 13 members labelled A, 1, 2, 3, 4, 5, 6, 7, 8, 9, 10, Q, K, and J. Each type of card has 3 picture cards labeled **Q**, **K** and **J**, called queen, king or jack. Hearts and diamonds are red while clubs and spades are black.

  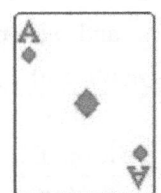 

Ace of hearts　　　Ace of clubs　　　Ace of diamonds　　　Ace of spades

Jack of hearts　　　Queen of clubs　　　King of diamonds　　　Jack of spades

---

 Example

1. A boy picks a card at random from a well-shuffled pack of 52 playing cards. What is the probability that
   (i) it is an Ace of heart　　　(ii) it is a king

   Solution
   $n(S) = 52$
   (i) $n(\text{Ace of hearts}) = 1$
   $\therefore P(\text{Ace of hearts}) = \dfrac{n(\text{Ace of hearts})}{n(S)} = \dfrac{1}{52}$
   (ii) $n(\text{Kings}) = 4$

125

$$\therefore P(\text{Kings}) = \frac{n(\text{Kings})}{n(S)} = \frac{4}{52} = \frac{1}{13}$$

2. Fourteen girls are sitting in a circle equally spaced. One is from form four, 2 are from form five, 6 are from lower sixth and 5 are from upper sixth. A girl is selected at random from amongst the girls. Find the probability that the girl is from

    (i) form four   (ii) form five   (iii) lower sixth   (iv) upper sixth

    Solution

    $n(S) = 14$, $n(\text{form four}) = 1$, $n(\text{form five}) = 2$

    $n(\text{Lower } 6^{th}) = 6$ and $n(\text{upper } 6^{th}) = 5$

    (i) $P(\text{form four}) = \dfrac{n(\text{form four})}{n(s)} = \dfrac{1}{14}$

    (ii) $P(\text{form five}) = \dfrac{n(\text{form five})}{n(5)} = \dfrac{2}{14} = \dfrac{1}{7}$

    (iii) $P(\text{lower } 6^{th}) = \dfrac{n(\text{lower } 6^{th})}{n(s)} = \dfrac{6}{14} = \dfrac{3}{7}$

    (iv) $P(\text{upper } 6^{th}) = \dfrac{n(\text{upper } 6^{th})}{n(s)} = \dfrac{5}{14}$

 Exercise 4:1

1. A die is tossed. Find the probability of:
    (a) obtaining a 2.   (b) obtaining an even number.
    (c) obtaining an odd number.   (d) obtaining a number less than 5.
    (e) obtaining a prime number.
2. The figure below shows a spinner. Find the probability of obtaining:
    (a) 3   (b) 1   (c) 2   (d) 5

3. In a race of twenty horses, 7 of the horses are black, 8 are white and the rest are dotted. Find the probability that the winner will be dotted.
4. At a conference, there are 9 boys, 12 girls, 15 men and 14 women. Find the probability that a person elected as president will be a man.
5. Find the probability of choosing at random the letter B from the letters of the word 'PROBABILITY'.
6. Find the probability of choosing a prime number at random from the integers 5 to 25 inclusive.

4.6 Complementary Events

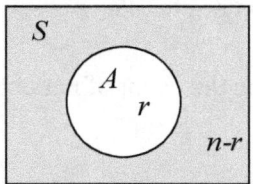

If A is an event in a sample space S, then the event that A does not occur is called "not A" or the "complement of A" denoted by $\bar{A}$ or A' and is defined as the union of all subsets of S whose elements do not belong to A. A is represented by the shaded portion in the Venn diagram above. Thus if the cardinality of S is n and the cardinality of A is r, then the cardinality of A' is given by

$$n(A') = n(s) - n(A)$$
$$n(A') = n - r$$

Therefore, if two events are complementary, the sum of their probabilities is 1.

$$P(A) + P(A') = 1 \Leftrightarrow P(A') = 1 - P(A)$$

Note!

A' is the complement of $A \Leftrightarrow A$ is the complement of A'.

Some examples of complementary events are:
1. The events "obtaining a head" and "obtaining a tail" when a coin is tossed.
2. The events "getting an even number" in one toss of die and "getting an odd number".

Example

1. The probability that a student will pass the GCE is $\dfrac{7}{11}$. What is the probability that he will fail?

   Solution
   Since the events, 'passing' and 'failing' are complementary,
   $P(\text{passing}) + P(\text{failing}) = 1$
   $$\Rightarrow P(\text{failing}) = 1 - P(\text{passing}) = 1 - \frac{7}{11} = \frac{4}{11}$$

2. A fair die is tossed. Find the probability of not be obtaining a two.

   Solution
   Let the event of obtaining a two be T, and the event of not obtaining a two be T'. Then,
   $$P(T) = \frac{1}{6}$$
   $$P(T') = 1 - P(T) = 1 - \frac{1}{6} = \frac{5}{6}$$

Exercise 4:2

   1. A bag contains 12 blue marbles, 14 red marbles and 9 green marbles. Find the probability that a marble drawn at random from the bag is
1. (a) green   (b) not blue   (c) red
2. A bag of mangoes contains 30 mangoes 8 of which are bad. A person takes mango at random from the bag. What is the probability that the mango is good?
3. A grocer bought 100 oranges. Given that, 16 of them are bad. Find the probability that a mango chosen at random from the lot is good.
4. In a class of 50 students, 35 are girls. Find the probability that a boy is the class prefect.
5. A poultry farm consists of 140 fowls 60 of which are hens. A man buys a fowl from the poultry; find the probability that it is a hen.
6. In a class of 80 students, 75 % of the students pass the examination. Find the probability that a student selected at random from the class failed.

4.7   Addition Laws of Probability (P (or))

We use the addition laws when the probability of one event **or** the other is required. Sometimes we use **either or** for emphasis. There are two cases involved. These include intersecting events and mutually exclusive events.

Intersecting Events

Consider any two events X and Y of an experiment which are such that $P(X) \neq 0$ and $P(Y) \neq 0$. We can represent X and Y as in the following Venn diagram.

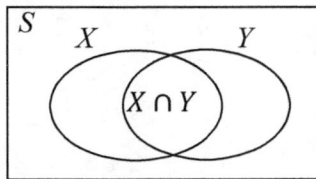

The probability of X or Y is given by

$$P(X \text{ or } Y) = P(X) + P(Y) - P(X \cap Y)$$

Since "X or Y" means, "only X occurs, or only Y occurs or both X and Y occur", we write X or Y in set notation as $X \cup Y$.

$$\therefore \ P(X \cup Y) = P(X) + P(Y) - P(X \cap Y)$$

 Example

1. In a group of 20 adults, 4 out of the 7 women and 2 out of the 13 men wear eye glasses. What is the probability that a person chosen at random from the group is:
    (i) a woman or someone who wears eye glasses?
    (ii) a man or someone who wears eye glasses?

    Solution
    Let M, W and G be the events of choosing a man, a woman and someone who wears eyeglasses respectively. Then,
    (i) $P(W \cup G) = P(W) + P(G) - P(W \cap G)$

    $$= \frac{7}{20} + \frac{6}{20} - \frac{4}{20} = \frac{9}{20}$$

    The probability that the person chosen is a woman or someone who wears glasses is $\dfrac{9}{20}$.

(ii) $P(M \cup G = P(M) + P(G) - P(M \cap G)$

$$= \dfrac{13}{20} + \dfrac{6}{20} - \dfrac{2}{20} = \dfrac{7}{20}$$

The probability that the person chosen is a man or someone who wears eyeglasses is $\dfrac{17}{20}$.

2. If X and Y are two events such that $P(X \cap Y) = \dfrac{1}{9}, P(X) = \dfrac{1}{3}$ and

$P(X \cup Y) = \dfrac{4}{9}$. Find $P(Y)$.

Solution

$$P(X \cup Y) = P(X) + P(Y) - P(X \cap Y)$$
$$\Rightarrow P(Y) = P(X \cup Y) + P(X \cap Y) - P(X)$$

$$= \dfrac{4}{29} + \dfrac{1}{9} - \dfrac{3}{9}$$

$$P(Y) = \dfrac{2}{9}$$

4.8 Mutually Exclusive Events

There are certain events for which the occurrence of one excludes the occurrence of the other. For instance, a tossed coin cannot land head up and tail up at the same time. We call such events mutually exclusive events. On a Venn diagram we represent them are as disjoint sets, since they cannot jointly occur at the same time. Hence, if two events X and Y are mutually exclusive,

$$X \cap Y = \varnothing \Leftrightarrow n(X \cap Y)$$
$$\therefore P(X \cap Y) = 0$$

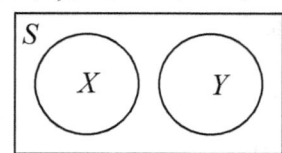

Therefore, the probability of one of them occurring is

$$P(X \text{ or } Y) = P(X \cup Y) = P(X) + P(Y)$$

In general if there are n mutually exclusive events $E_1, E_2,...,E_n$, then the occurrence of one excludes the occurrence of all the others. Hence,

$$P(E_1 \cup E_2 \cup ... \cup E_n) = P(E_1) + P(E_2) + \cdots + P(E_n)$$

 Example

1. In an election for the president of a credit union, the probability that Shey (S) wins is 0.3, the probability that Chia (C) wins is 0.2 and the probability that Doh (D) wins is 0.4. Find the probability that:
   (a) Shey or Chia wins.   (b) neither Shey nor Doh wins.

   Solution
   The events are mutually exclusive, so
   (a) $P(S \text{ or } C) = P(S) + P(C) = 0.3 + 0.2 = 0.5$
   (a) $P(\text{neither } S \text{ nor } D) = 1 - \{P(S) + P(D)\} = 1 - (0.3 + 0.4) = 0.3$

2. Find the probability that a card drawn at random from an ordinary deck of 52 cards is either a king or an Ace of spade.

   Solution
   Let A and K be the events drawing an Ace of spade and a king respectively. Then,

   $$P(A) = \frac{n(A)}{n(S)} = \frac{1}{52}$$

   $$P(K) = \frac{n(K)}{n(S)} = \frac{4}{52}$$

   $$P(A \cup K) = P(A) + P(K) = \frac{1}{52} + \frac{4}{52} = \frac{5}{52}$$

3. A coin is tossed twice. Find the probability that the result is either two tails or a tail and a head respectively.

   Solution
   $$S = \{HH, HT, TT, TH\}$$

   $$P(TT) = \frac{1}{4}, P(TH) = \frac{1}{4}$$

   $$\therefore P(TT \cup TH) = P(TT) + P(TH) = \frac{1}{4} + \frac{1}{4} = \frac{1}{2}$$

Exercise 4:4

1. There are 12 cards numbered 1-12. Find the probability that a card selected at random from the cards is even or a perfect square.
2. Two events A and B are such that
   $P(A) = \frac{1}{2}, P(B) = \frac{1}{2}$ and $P(A \cap B) = \frac{1}{3}$. Calculate the probability of A or B occurring.
3. Find the probability that the sum will be 7 or 11 if we throw two fair dice at the same time.
4. A bag contains 5 red, 7 blue and 6 green marbles. What is the probability that a boy with closed eyes picks a green or red marble from it?
5. A number is chosen at random from the set $\{15, 16, 17,\ldots, 32\}$. Find the probability that the chosen number is:
   (i) a multiple of 3
   (ii) a prime number
   (iii) a multiple of 3 or a prime number
6. A bucket contains 6 mangoes, 11 bananas and 13 oranges. What is the probability that a fruit chosen at random from the bucket is either a mango or a banana?
7. A bag contains 2 red, 4 green and 3 white balls. Find the probability that a ball selected at random from the bag is either red or green.
8. A woman is to select a day convenient for her to attend a meeting in the Town Hall. What is the probability that she will choose:
   (a) a day that begins with the letter T.
   (b) a day that begins with the letter T or M.
9. Find the probability that a letter chosen at random from the 26 letters of the English alphabet is:
   (a) Either letter f or j.
   (b) One of the letters of the word 'TRIANGLE'
10. Find the probability that a number chosen at random from the numbers 1 to 10 is:
    (a) a prime number   (b) a multiple of 3
    (c) a prime number or a multiple of 3.
11. Find the probability that if we toss a dice once we will obtain:
    (a) 5   (b) 2   (c) 5 or 2
12. Find the probability of obtaining 5 or 4 in one toss of a fair cubic die.
13. The probability that candidate A, wins an election is $\frac{4}{9}$, the probability that candidate B, wins is $\frac{2}{5}$ and the probability that candidate C, wins is $\frac{3}{7}$. Find the probability that:
    (a) A or B wins   (b) B or C wins   (c) Neither A nor C wins.

14. There are 40 students in a class. All students offer Further Mathematics (*F*) or Human Biology (*H*). 27 offer Further Mathematics, and 24 offer Human Biology. Find the probability that a student chosen at random from the class offers:
    (a) Further Mathematics.
    (b) Human Biology.
    (c) Further Mathematics or Human Biology.
15. Given that $P(A) = \frac{3}{5}, P(B) = \frac{4}{7}$ and $P(A \cup B) = \frac{5}{8}$, find $P(A \cap B)$.

4.9   The Multiplication Law

The multiplication laws are used when the probability of one event **and** the other or **both** events is required. There are two cases involved-dependent events and independent events.

4.10 Dependent Events

If two events *X* and *Y* are such that $P(X) \neq 0$ and $P(Y) \neq 0$, then the probability of *X* given that *Y* has already occurred is denoted by $P(X/Y)$ and is given by

$$P(X/Y) = \frac{n(X \cap Y)}{n(Y)} = \frac{P(X \cap Y)}{P(Y)}$$

$$\Rightarrow P(X \cap Y) = P(Y) \cdot P(X/Y)$$

This is known as the multiplication law for dependent events

 Example

Given that a heart is picked at random from a pack of 52 playing cards, find the probability that the next card chosen is a picture card.

Solution

$$P(P/H) = \frac{P(P \cap H)}{P(H)} = \frac{3}{52} \div \frac{13}{52} = \frac{3}{13}$$

Alternatively:

$$P(P/H) = \frac{\text{number of heart picture cards}}{\text{number of hearts}} = \frac{3}{13}$$

4.11 Independent Events

Consider the two events "a student eats in the morning" and "the teacher of the student is rich"

Clearly, the event that a student eats in the morning does not in any way depend or affect the event that his teacher is rich. If two events X and Y are such that the occurrence or the non-occurrence of X does not in any way affect or depend on the occurrence or non-occurrence of Y, then the events X and Y are said to be **independent**.

If two events X and Y are independent then,

$$P(X/Y) = P(X) \text{ and } P(Y/X) = P(Y)$$

$$\text{but } P(X/Y) = \frac{P(X \cap Y)}{P(Y)}$$

$$\Rightarrow P(X) = \frac{P(X \cap Y)}{P(Y)}$$

$$\Rightarrow P(X \cap Y) = P(X) \cdot P(Y)$$

We call this law the **multiplication law for independent events.**

For n independent events

$$P(E_1 \cap E_2 \cap \dots \cap E_n) = P(E_1) \times P(E_2) \times \dots \times P(E_n)$$

 Example

1. Find the probability that a die and a coin thrown in succession will show a head on the die and a two on the coin.

   Solution

   $$P(2 \cap H) = P(2) \cdot P(H) = \frac{1}{6} \cdot \frac{1}{2} = \frac{1}{12}$$

2. Find the probability that two coins tossed simultaneously turn up heads if one of

the coins is fair and the other is twice as likely to turn up a head as a tail.

Solution

Let H_f and H_u represent the events that a head is shown on the fair and unfair coins respectively then $P(H_f) = \dfrac{1}{2}$ and $P(H_u) = \dfrac{2}{3}$

Since these are independent events

$$P(H_f \cap H_u) = P(H_f) \cdot P(H_u) = \frac{1}{2} \cdot \frac{2}{3} = \frac{1}{3}$$

3. Find the probability that on rolling a fair die numbered 1 to 6 twice,
    (i)   We obtain two sixes.
    (ii) The second throw will be a six given that the first throw is a six.
    (iii) We obtain a score of ten from the two throws.
    (iv) We obtain at least one six.
    (v)   We obtain exactly one six.

Solution

(i) $P(6) = \dfrac{1}{6} \Rightarrow P(2 \text{ sixes}) = P(6 \cap 6)$

$$= \left(\frac{1}{6}\right)\left(\frac{1}{6}\right) = \frac{1}{36}$$

(ii) $P(6/6) = \dfrac{P(6 \cap 6)}{P(6)} = \dfrac{1}{36} \div \dfrac{1}{6} = \dfrac{1}{36} \times \dfrac{6}{1} = \dfrac{1}{6}$

(iii) $P(\text{score of ten}) = P(4.6) + P(6,5) + P(5,5)$

$$= \frac{1}{36} + \frac{1}{36} + \frac{1}{36} = \frac{1}{12}$$

(iv) $P(\text{at least one six}) = \dfrac{11}{36}$ (v) $P(\text{exactly one six}) = \dfrac{10}{36} = \dfrac{5}{18}$

4.12 Conditional Probability

Drawing With and Without Replacement

 Example

1. A bag contains 7 red balls and 3 white balls.
    Find the probability that a ball chosen at random from the bag is
    (i) red   (ii) white.

Solution

Let the events "choosing a red ball" and "choosing a white ball" be R and W respectively. Then,

(i) $P(R) = \dfrac{7}{10}$ (ii) $P(W) = \dfrac{3}{10}$

2. Suppose that in Example 1 above, we choose red ball and do not replace it.
    (i)   How many red balls are remaining?
    (ii)   How many balls are remaining altogether?
    (iii) Find the probability that the second ball chosen at random from the bag is
           (a) red.   (b) white.
    (iv)   Suppose the second ball we choose was red again and we did not replace it but added 3 white balls to the bag. What will be the probability that a third ball chosen at random from the bag will be
           (a) red.   (b) white.

Solution
(i) $n(R) = 6$ (ii) $n(S) = 9$

(iii) (a) $P(R) = \dfrac{6}{9} = \dfrac{2}{3}$ (b) $P(W) = \dfrac{3}{9} = \dfrac{1}{3}$

(iv) $n(R) = 5$ (ii) $n(W) = 6$

      (a) $P(R) = \dfrac{5}{11}$ (b) $P(W) = \dfrac{6}{11}$

4.13 Repeated Trials

For repeated experiments, if the conditions of the experiments are unaltered, we consider that the events are independent.

 Example

Find the probability of having a head, a head, a tail, a head, and a tail in that order on tossing biased coin, if $P(H) = 2P(T)$.

Solution

$P(H) + P(T) = 1 \Longrightarrow 2P(T) + P(T) = 1 \Longrightarrow 3P(T) = 1$ and $P(T) = \dfrac{1}{3}$

$$P(HHTHT) = P(H).P(H).P(T).P(H).P(T)$$

$$= \left(2P(T)\right)^3 \left(P(T)\right)^2 = 8\left(\frac{1}{3}\right)^5 = \frac{8}{243}$$

4.14 Tree Diagrams

Tree diagrams are devices used to enumerate all the possible outcomes of a sequence of experiments, when each experiment occurs in a finite number of ways in solving probability problems especially when the problem consists of a mixture of mutually exclusive events and independent events.

 Example

1. Draw a tree diagram to represent the possible outcomes of tossing a fair coin twice. Hence, find the probability of obtaining:
   (a) two heads   (b) a head and a tail.

   Solution

   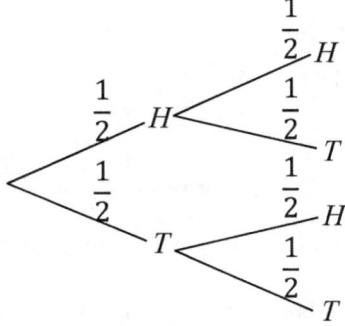

   (a) $P(HH) = \frac{1}{2}\left(\frac{1}{2}\right) = \frac{1}{4}$

   (b) $P\left(HT \text{ or } TH\right) = \left(\frac{1}{2}\right)\left(\frac{1}{2}\right) + \left(\frac{1}{2}\right)\left(\frac{1}{2}\right) = \frac{1}{2}$

2. A game consists of tossing a coin. If a head shows, we throw a fair die once and note the score. If a tail shows, we select a card from a well-shuffled pack of 52 playing cards. Hearts score 1, diamonds score 2, clubs score 3 and spades score

137

4. the scores 1,2,3,4,5,6 win 100 frs, 200 frs, 300 frs, 400 frs, 500 frs, 600 frs respectively. Draw a tree diagram to show all the possible outcomes for a single throw of a coin and find the probability of winning 300 frs or more. To register for the game one needs 400 frs. Would you advice your friend to play the game?

Solution

For convenience

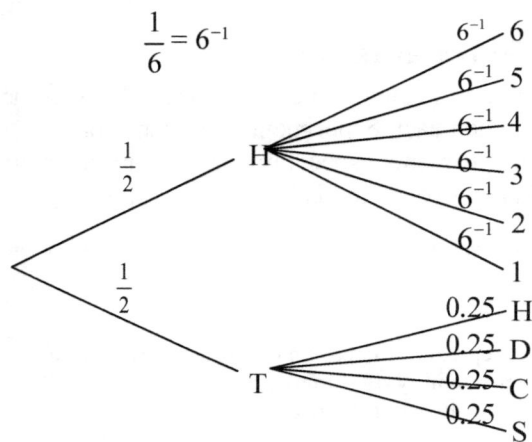

$$\frac{1}{6} = 6^{-1}$$

Probability of winning 300 frs or more = $P(\text{Score} \geq 3)$

$P(\text{Score} \geq 3) = 1 - P(\text{Score} \leq 3)$

$P(\text{Score} \leq 3) = 3\left(\frac{1}{2}\right)\left(\frac{1}{6}\right) + 3\left(\frac{1}{2}\right)\left(\frac{1}{4}\right) = \frac{2}{8} + \frac{3}{8} = \frac{5}{8}$

$P(\text{Score} \geq 3) = 1 - \frac{5}{8} = \frac{3}{8}$.

Therefore, probability of winning 300 frs or more is $\frac{3}{8}$.

The game is very unfavourable so I will not advice my friend to play.

3. A coin is tossed 4 times. Find the probability of obtaining 3 heads and one tail.

Solution

Let $P_{3,1}$ = Probability of 3 heads and one tail

$P_{3,1} = P(\text{HHHT}) + P(\text{HHTH}) + P(\text{HTHH}) + P(\text{THHH})$

$$P(\text{HHHT}) = \left(\frac{1}{2}\right)\left(\frac{1}{2}\right)\left(\frac{1}{2}\right)\left(\frac{1}{2}\right) = \frac{1}{16}$$

$$P(\text{HHTH}) = \left(\frac{1}{2}\right)\left(\frac{1}{2}\right)\left(\frac{1}{2}\right)\left(\frac{1}{2}\right) = \frac{1}{16}$$

$$P(HTHH) = \left(\frac{1}{2}\right)\left(\frac{1}{2}\right)\left(\frac{1}{2}\right)\left(\frac{1}{2}\right) = \frac{1}{16}$$

$$P(THHH) = \left(\frac{1}{2}\right)\left(\frac{1}{2}\right)\left(\frac{1}{2}\right)\left(\frac{1}{2}\right) = \frac{1}{16}$$

$\therefore$ Probability of 3 heads and 1 tail $= 4\left(\frac{1}{16}\right) = \frac{1}{4}$.

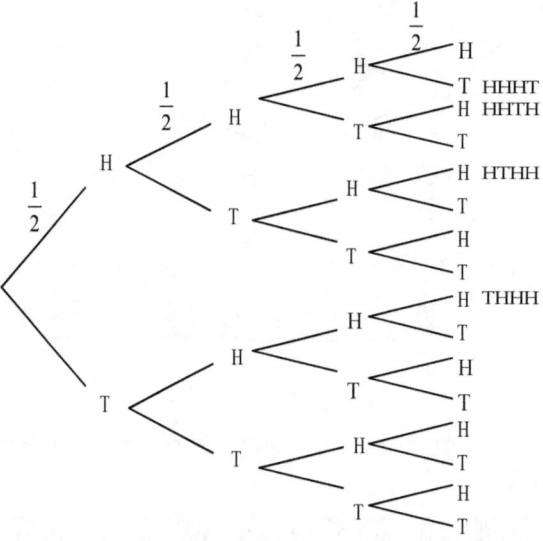

4. Find the probability of rolling exactly two sixes using three dice each numbered 1 to 6, given that one die is fair and the others are biased so that for each of these a six is twice as likely as any other score.

Solution

Let the event of obtaining a six be S, then the event of obtaining no six is $\overline{S}$.

On the fair die $P(S) = \frac{1}{6}$ and $P(\overline{S}) = \frac{5}{6}$

Let x be the probability of obtaining 1, 2, 3, 4 or 5 on the unfair die and $P(S)$ be probability of obtaining a 6.

Then, $5x + 2x = 1 \Rightarrow x = \frac{1}{7}$.

On the unfair die, $P(S) = \frac{2}{7}$ and $P(\overline{S}) = \frac{5}{7}$.

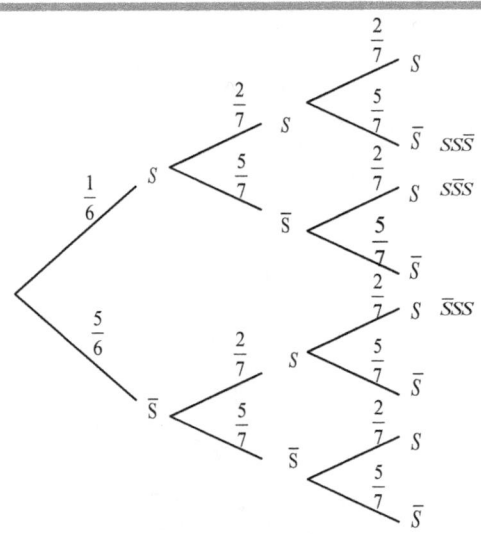

$$P(\text{exactly 2 sixes}) = \left(\frac{1}{6}\right)\left(\frac{2}{7}\right)\left(\frac{5}{7}\right) + \left(\frac{1}{6}\right)\left(\frac{5}{7}\right)\left(\frac{2}{7}\right) + \left(\frac{5}{6}\right)\left(\frac{2}{7}\right)\left(\frac{2}{7}\right) = \frac{20}{147}$$

 Exercise 4:5

1. A bag contains 3 black balls and 2 red balls. We pick a ball at random from the bag and then replaced it before drawing a second ball. What is the probability that
    (a) They are both black.   (b) they are both red.
    (c) One is black and the other is red.
2. Find the probability of scoring a two on one die and a five on another on rolling a pair of fair dice with each face numbered 1-6.
3. A student has 3 khaki shirts and 2 white shirts; he has also 3 pairs of khaki trousers and 4 pairs of white trousers. One night in the dark, he chooses one shirt and one pair of trousers at random. Find the probability that the shirt and the trousers are of
    (a) The same colour. (b) Different colour.
4. A bag contains 5 white socks and 6 blue socks. Find the probability that if two socks are drawn at random in succession and without replacement, they will be
    (a) Both white.   (b) Both blue.
    (c) Of the same colour   (d) of different colour
5. There are four children in a certain family. Find the probability that the

family has:
(a) An equal number of boys and girls. (b) At least two boys.
(c) More boys than girls (d) Children of the same sex.

6. Find the probability of drawing two red balls from a box containing two red balls and four blue balls given that we draw a ball at random from the box and replace it before drawing a second ball.

7. A staff consists of 15 men and 10 women. 60 % of the men and 40% of the women can drive. Find the probability that if we choose man and a woman at random from the staff,
(a) Both of them can drive. (b) Only the man can drive.
(c) Only the woman can drive. (d) None of them can drive.
(e) Only one of them can drive.

8. A bag contains 9 identical balls out of which 3 are blue, 2 are white and the rest are red. If two balls are drawn at random one after the other without replacement, what is the probability that:
(a) Both of them will be blue? (b) They will be of the same colour?
(c) They will be of different colour?

9. A packet contains four red, five blue and six black identical balls. Find the probability of picking if we pick two balls at random from the bag without replacement,
(a) One will be red and the other black.
(b) The balls will be of the same colour.
(c) The balls will be of different colour.

10. Find the probability of getting at least one tail on tossing two fair coins at once.

11. A pair of fair dice each numbered 1 to 6 are tossed. Find the probability of obtaining a sum of at least nine.

12. The probability that a civil servant owns a car is $\frac{3}{7}$. Find the probability that:
(a) Two civil servants selected at random each own a car.
(b) Of two civil servants selected at random, only one owns a car.
(c) Of three civil servants selected at random, only one owns a car.

13. Find the probability of choosing at random a multiple of five or three from the integers 5 to 25 inclusive.

14. A bag contains 10 identical balls of which 4 are blue and 6 are red. Find the probability that if we pick two balls at random one after the other with replacement,
(a) Both balls will be red. (b) Both will be of the same colour.

15. Find the probability of obtaining a total score of six on throwing two perfect dice once.

16. A farmer added 6 more ripe mangoes to a basket containing 24 mangoes, of which m are ripe.
(a) Find in terms of m, the number of unripe mangoes in the basket.

(b) Find in terms of *m*, the probability that a mango chosen at random from the basket is ripe.

17. The probability of selecting a green ball from a bag containing six more green balls than red balls is $\frac{2}{3}$. Find the number of balls in the bag if it contains just red and green balls.

18. Find the probability that a card drawn at random from a pack of 52 well-shuffled cards is:
    (a) a queen of clubs or hearts.   (b) a red card but not a queen.

4.15 Probability from Frequency Tables

 Example

1. The following table shows the scores out of 100 obtained by 36 students in a test.

Marks	f
0-9	0
10-19	1
20-29	2
30-39	4
40-49	7
50-59	9
60-69	5
70-79	3
80-89	3
90-99	2

(a) Make a cumulative frequency table for the data.
(b) Hence, calculate the probability that a student chosen at random from the class scored a mark:
    (i)   Less than 69.5%.   (ii) Greater than or equal to 49%.
    (iii) From 30% to 79% inclusively.

Solution

(a)

Mark	Cumulative
< 9.5	0
< 19.5	1
< 29.5	3
< 39.5	7
< 49.5	14
< 59.5	23
< 69.5	28
< 79.5	31
< 89.5	34
< 99.5	36

(b) (i) $P(X < 69.5) = \dfrac{28}{36} = \dfrac{7}{9}$

(ii) $P(X \geq 49.5) = 1 - P(X < 49.5)$

$\Rightarrow P(X \geq 49.5) = 1 - \dfrac{14}{36} = \dfrac{11}{18}$

(iii) $P(30 \leq X \leq 79) = \dfrac{31 - 3}{36} = \dfrac{7}{9}$

Exercise 4:6

1. A Mathematics Teacher drives to school every day. She records her journey time, to the nearest minutes, over a period of 60 days as follows:

Time (min)	8	9	10	11	12	13	14	15
No. of days	5	7	13	12	11	7	3	2

(a) State the modal time.

(b) Calculate her mean journey time.

(c) Estimate, to 2 decimal places, the standard deviation. Find the probability that:

(d) She took 11 minutes to get to school on a certain day.

(e) She was not late on a certain day, given that she left her house 10 minutes before the start of her lesson.

2. The masses of 100 people were recorded to the nearest kg and displayed as follows:

Mass (kg)	Number of people
10-19	6
20-29	15
30-39	39
40-49	22
50-59	12
60-69	5
70-79	1

(a) Determine the width of the class intervals.
(b) State the modal class.
(c) Calculate an estimate of the mean mass of the students to the nearest of 10th of a kg.
(d) Using the data draw a cumulative frequency curve.
(e) Find the inter-quartile range.
(f) Find the probability that a student chosen at random will weigh less than 55 kg.

3. In a research institute the approximate weights of 500 oranges were recorded as follows;

Weights (g)	Number of oranges
0-29	20
30-59	25
60-89	50
90-119	75
120-149	120
150-179	100
180-209	80
210-239	30

(i) State the modal class of the distribution.
(ii) Estimate to one decimal place the mean of the distribution.
(iii) Construct a cumulative frequency table for the distribution. Taking 1cm to represent 25 oranges and 1 cm to represent 15 g draw a cumulative frequency graph to represent this data. Use your graph to estimate:
(a) The median of the distribution.
(b) The number of oranges, which weigh less than 75 g each.
(c) The probability that an orange selected at random will have a weight greater than or equal to 175 g.

Multiple Choice Exercise 4

1. There are m boys and 12 girls in class. The probability of selecting at random a girl from the class is:

   [A] $\dfrac{m}{12}$ [B] $\dfrac{12}{m}$ [C] $\dfrac{12}{m+12}$ [D] $\dfrac{m}{m+12}$

2. The table below gives the marks scored by a group of students in a test. The probability of selecting a student from the group that scored 2 or 3 is:

   [A] $\dfrac{1}{11}$ [B] $\dfrac{5}{25}$ [C] $\dfrac{7}{22}$ [D] $\dfrac{6}{11}$

Mark	0	1	2	3	4	5
Frequency	1	2	7	5	4	3

3. The probability of having an odd number in a single toss of a fair die is:

   [A] $\dfrac{2}{3}$ [B] $\dfrac{1}{6}$ [C] $\dfrac{1}{3}$ [D] $\dfrac{1}{2}$

4. The table below gives the scores of a group of students in an English Language test. The probability that a student chosen at random from the group scored at least six marks is:

   [A] $\dfrac{3}{4}$ [B] $\dfrac{1}{5}$ [C] $\dfrac{1}{4}$ [D] $\dfrac{3}{10}$

Score	2	3	4	5	6	7
Number of students	2	4	7	2	3	2

5. The table below shows the results of two groups of students who cast their votes on a particular proposal. The probability that if we select a student for a post in favour of the proposal, he is from group A is:

   [A] $\dfrac{8}{9}$ [B] $\dfrac{16}{35}$ [C] $\dfrac{4}{5}$ [D] $\dfrac{4}{7}$

	In favour	Against
Group A	128	32
Group B	96	48

6. Two groups of students cast their votes on a particular proposal. The results are as in Table 38:8. If we choose a student at random, the probability that he is against the proposal is:

   [A] $\dfrac{3}{19}$ [B] $\dfrac{4}{19}$ [C] $\dfrac{5}{19}$ [D] $\dfrac{9}{19}$

7. The events X and Y are mutually exclusive and $P(X) = \dfrac{1}{3}$, $P(Y) = \dfrac{2}{5}$.

    $P(X \cap Y)$ is:

    [A] 0   [B] $\dfrac{2}{15}$ [C] $\dfrac{4}{15}$ [D] $\dfrac{11}{15}$

8. The events X and Y are mutually exclusive and $P(X) = \dfrac{1}{3}$, $P(Y) = \dfrac{2}{5}$.

    $P(X \cup Y)$ is:

    [A] 0   [B] $\dfrac{2}{15}$ [C] $\dfrac{4}{15}$ [D] $\dfrac{11}{15}$

9. A box contains two white and three blue identical marbles. The probability of picking at random one after the other without replacement two marbles of different colours is:

    [A] $\dfrac{2}{3}$ [B] $\dfrac{3}{5}$ [C] $\dfrac{2}{5}$ [D] $\dfrac{3}{10}$

10. Mrs. Ngala is expecting a baby. The probability of a boy is $\dfrac{1}{2}$ and the

    probability that the baby will have blue eyes is $\dfrac{1}{4}$. The probability that she

    will have a blue-eyed boy is:

    [A] $\dfrac{1}{8}$ [B] $\dfrac{1}{4}$ [C] $\dfrac{3}{8}$ [D] $\dfrac{3}{4}$

11. A number is chosen at random from the set $\{1,2,3,\cdots,9,10\}$. The probability that the number is greater than or equal to seven is:

    [A] $\dfrac{1}{10}$ [B] $\dfrac{3}{10}$ [C] $\dfrac{2}{5}$ [D] $\dfrac{3}{5}$

12. The probability of throwing a number greater than 2 with a single fair die is:

    [A] $\dfrac{1}{6}$ [B] $\dfrac{1}{3}$ [C] $\dfrac{1}{2}$ [D] $\dfrac{2}{3}$

13. The probability of obtaining 4 or 6 on rolling a fair die once is:

    [A] $\dfrac{2}{3}$ [B] $\dfrac{1}{6}$ [C] $\dfrac{1}{3}$ [D] $\dfrac{1}{2}$

14. Three balls are drawn one after the other with replacement, from a bag containing five red, nine white and four blue identical balls. The probability that they are one red, one white and one blue is:

    [A] $\dfrac{5}{81}$ [B] $\dfrac{5}{27}$ [C] $\dfrac{5}{162}$ [D] $\dfrac{5}{243}$

15. The probability that an integer selected from the set of integers $\{20,21,\cdots,30\}$ is a prime number is:

[A] $\frac{2}{11}$ [B] $\frac{5}{11}$ [C] $\frac{6}{11}$ [D] $\frac{9}{11}$

16. The probability of obtaining a number less than three is on rolling a fair die once is:

[A] $\frac{1}{6}$ [B] $\frac{1}{3}$ [C] $\frac{1}{2}$ [D] $\frac{2}{3}$

17-18. The data below shows the number of workers employed in the various sections of a construction company in Yaounde. Use the information to answer questions 17 to 18. 24 Carpenters, 27 Labourers, 12 Plumbers, 15 Plasterers, 9 Painters, 3 Messengers and 18 Bricklayers.

17. One of the workers is absent on a day. The probability that he is a bricklayer is:

[A] $\frac{1}{9}$ [B] $\frac{2}{9}$ [C] $\frac{1}{6}$ [D] $\frac{1}{4}$

18. The probability that a worker retrenched is a plumber or a plasterer is:

[A] $\frac{3}{4}$ [B] $\frac{1}{9}$ [C] $\frac{5}{36}$ [D] $\frac{1}{4}$

19. The probability that a total sum of seven would appear with two tosses of a fair die is:

[A] $\frac{5}{36}$ [B] $\frac{1}{6}$ [C] $\frac{7}{36}$ [D] $\frac{5}{6}$

20. A die is rolled 200 times. The outcomes obtained are shown in Table 38:9. The probability of obtaining a 2 is:
[A] 0.002 [B] 0.015 [C] 0.15 [D] 0.16

21. A die is rolled 200 times. The outcomes obtained are shown in the table below. The probability of obtaining a number less than 3 is:
[A] 0.125 [B] 0.150 [C] 0.275 [D] 0.500

Number	1	2	3	4	5	6
Number of times	25	30	45	28	40	32

22. Two cards are drawn one after the other with replacement from a well shuffled ordinary deck of 52 cards containing four aces. The probability that they are both aces is:

[A] $\frac{1}{13}$ [B] $\frac{1}{169}$ [C] $\frac{1}{52}$ [D] $\frac{1}{26}$

23. The probability that a number selected from the numbers 30 to 50 inclusive is a prime is:

[A] $\frac{1}{4}$ [B] $\frac{5}{21}$ [C] $\frac{3}{7}$ [D] $\frac{1}{3}$

24. Two fair dice are tossed together once. The probability that the sum of the outcome is at least ten is:

[A] $\dfrac{1}{12}$ [B] $\dfrac{5}{36}$ [C] $\dfrac{1}{6}$ [D] $\dfrac{5}{18}$

25. From a box containing 2 red, 6 white and 5 blackballs, a ball is randomly selected. The probability that the ball selected is black is:

[A] $\dfrac{2}{13}$ [B] $\dfrac{5}{13}$ [C] $\dfrac{5}{11}$ [D] $\dfrac{11}{13}$

26. A bag contains three red, four black and five green identical balls. The probability that two balls picked at random one after the other without replacement will be one red and the other green is:

[A] $\dfrac{5}{48}$ [B] $\dfrac{5}{11}$ [C] $\dfrac{5}{22}$ [D] $\dfrac{5}{44}$

27. The table below gives the distribution of outcomes obtained when a die was roll 100 times. The experimental probability that it shows at most four when rolled again is:

[A] $\dfrac{8}{25}$ [B] $\dfrac{12}{25}$ [C] $\dfrac{13}{25}$ [D] $\dfrac{17}{25}$

Number of die	1	2	3	4	5	6
Frequency	18	14	20	16	15	17

28. A bag contains red, black, and green identical balls. We pick a ball and replace it. The table below shows the result of 100 trials. The experimental probability of picking a green ball is:

Colour of ball	red	black	green
Number of occurrences	54	30	16

[A] $\dfrac{4}{25}$ [B] $\dfrac{21}{25}$ [C] $\dfrac{1}{3}$ [D] $\dfrac{4}{21}$

29. A box contains two white and three blue identical balls. We pick two balls at random one after the other, without replacement. The probability of picking two balls of different colours is:

[A] $\dfrac{6}{25}$ [B] $\dfrac{7}{20}$ [C] $\dfrac{3}{5}$ [D] $\dfrac{2}{3}$

30. From a group of eleven people, seven can speak English and six can speak French. The probability that a person chosen at random can speak both English and French is:

[A] $\dfrac{2}{11}$ [B] $\dfrac{4}{11}$ [C] $\dfrac{5}{11}$ [D] $\dfrac{11}{13}$

31. The probabilities that Awah and Suh pass an examination are $\dfrac{3}{4}$ and $\dfrac{3}{5}$

respectively. The probability of both boys failing the examination is:

[A] $\frac{2}{3}$ [B] $\frac{3}{10}$ [C] $\frac{9}{10}$ [D] $\frac{1}{10}$

32. A box contains five red, three green and four blue balls. A boy takes away two balls at random from the box. The probability that the two balls are red is:

[A] $\frac{5}{33}$ [B] $\frac{5}{36}$ [C] $\frac{103}{132}$ [D] $\frac{31}{36}$

33. A box contains five red, three green and four blue balls. A boy takes away two balls at random from the box. The probability that one is green and the other is blue is:

[A] $\frac{2}{11}$ [B] $\frac{5}{12}$ [C] $\frac{8}{12}$ [D] $\frac{7}{11}$

34. The probability that an event will occur is p and the probability that it will not occur is q. The true assertion is:

[A] $p-q=1$ [B] $q-p=0$ [C] $p+q=1$ [D] $p+q=0$

35. The probability of selecting a prime number at random from the set $Y = \{18, 19, 20, \cdots, 28, 29\}$ is:

[A] $\frac{1}{4}$ [B] $\frac{3}{11}$ [C] $\frac{1}{2}$ [D] $\frac{3}{4}$

36. The numbers of goals scored by a school team in 10 netball matches are: 3,5,7,7,8,8,8,11,11,12. The probability that in a match, the school team will score at most 8 goals is:

[A] $\frac{1}{5}$ [B] $\frac{2}{5}$ [C] $\frac{3}{5}$ [D] $\frac{7}{10}$

37. The probability that a number chosen at random from {2,3,4,...,9,10} is either a prime number or a multiple of 3 is:

[A] $\frac{5}{9}$ [B] $\frac{2}{3}$ [C] $\frac{6}{7}$ [D] $\frac{5}{7}$

38. The probabilities that Ade and his dog will be alive in 10 years time are 0.8 and 0.3 respectively. The probability that they will both be alive in 10 years time is:

[A] 1.00   [B] 0.50   [C] 0.24   [D] 0.06

39. The probabilities of Fru and Nsang passing an examination are $\frac{3}{4}$ and $\frac{5}{8}$ respectively. The probability that the two boys fail the examination is:

[A] $\frac{3}{32}$ [B] $\frac{3}{8}$ [C] $\frac{15}{32}$ [D] $\frac{5}{8}$

40. Two beads are drawn at random, one after the other with replacement, from a box containing 5 red and 7 white identical beads. The probability that the beads are the same colour is:

[A] $\dfrac{119}{144}$ [B] $\dfrac{95}{144}$ [C] $\dfrac{37}{72}$ [D] $\dfrac{49}{144}$

41. The probability that John passes the GCE is $\dfrac{2}{3}$ and the probability that Paul fails the same exam is $\dfrac{1}{4}$. The probability that both John and Paul pass is:

[A] $\dfrac{1}{6}$ [B] $\dfrac{1}{2}$ [C] $\dfrac{3}{4}$ [D] $\dfrac{5}{12}$

42. The probability of obtaining exactly one head on tossing a coin twice is:

[A] $\dfrac{1}{4}$ [B] $\dfrac{1}{2}$ [C] $\dfrac{3}{4}$ [D] 1

Module 20

Solid Figures

Family of Situations

Module 20 is an extension of module 3, 7 and 12 at the end of the module; the student is expected to acquire many more competencies within the **families of situations** *'Usage of Technical Objects in everyday life'*.

Categories of Action

The categories of action for module 20 include:
1. Recognition of Objects,
2. Production of Objects,
3. Determination of Measures and how much an object can contain.

Credit

The module is expected to be covered within 5 weeks teaching 4 hours per week (or within 20 hours).

Topic 5

Mensuration of Solids Figures

Objectives

At the end of this topic, the learner should be able to:

1. Observe and describe a prism
2. Recognize and identify a right prisms
3. Identify the apex, lateral surface, lateral edge, altitude of a prism.
4. Make sketches of prisms.
5. Draw and make net of prisms
6. Make models of prisms from nets.
7. Use the various parts of the net to establish the original figure.
8. Calculate total surface area and volume of prisms.
9. Calculate total surface area and volume of cubes.
10. Calculate total surface area and volume of cylinders.

5.1   Properties of Solids

We did much work on solid figures in modules 3, 7 and 12. The learner may go back and revise book 1, topics 14 and 15; book 2, topics 9, 10 and 11, and book 3, topic 7 before continuing.

The following is an extract of the formulae for finding the volume and surface area of some popular solids.

Diagram	Name of solid	Volume	Surface area
	Cuboid	$V = lwh$	$2lw + 2hw = 2lh$
	cylinder	$V = \pi r^2 h$	$S = 2\pi rh + 2\pi r^2$
	Cone	$V = \dfrac{1}{3}\pi r^2 h$	$S = \dfrac{1}{2}(2\pi r)l + \pi r^2$ $S = \pi rl + \pi r^2$
	Square pyramid	$V = \dfrac{1}{3}l^2 h$	$S = l^2 + 4\left(\dfrac{1}{2}lh\right)$ $S = l^2 + 2lh$
	sphere	$V = \dfrac{4}{3}\pi r^3$	$S = 4\pi r^2$

Review and Revision Exercise

(1) State the most appropriate name of each of the following solids.

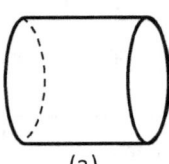

(a)

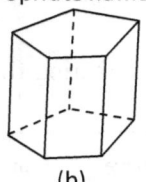

(b)

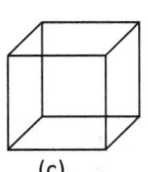

(c)

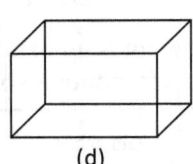

(d)

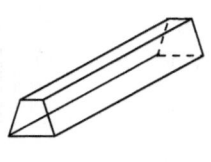

(e)

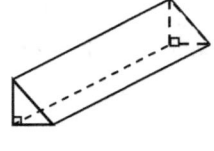

(f)

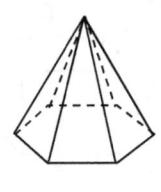

(g)

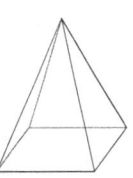

(h)

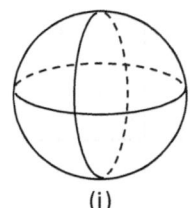

(i)

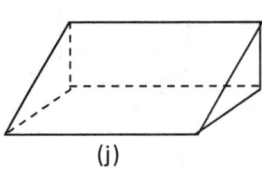

(j)

(k)

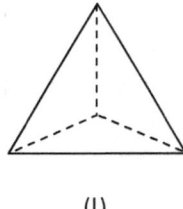

(l)

(m)

(2) Describe each of the solids in (1) above.

(3) Draw a diagram of each of the solids named below.
   (a) a hexagonal pyramid   (b) a sphere   (c) a cube
   (d) a frustum   (e) a cylinder   (f) a cone
   (g) a trapezoidal prism   (h) a tetrahedron   (i) a hexagonal pyramid
   (j) a pentagonal prism   (k) a cuboid   (l) a square pyramid

(4) Name the solid that can be formed from each of the nets below.

154

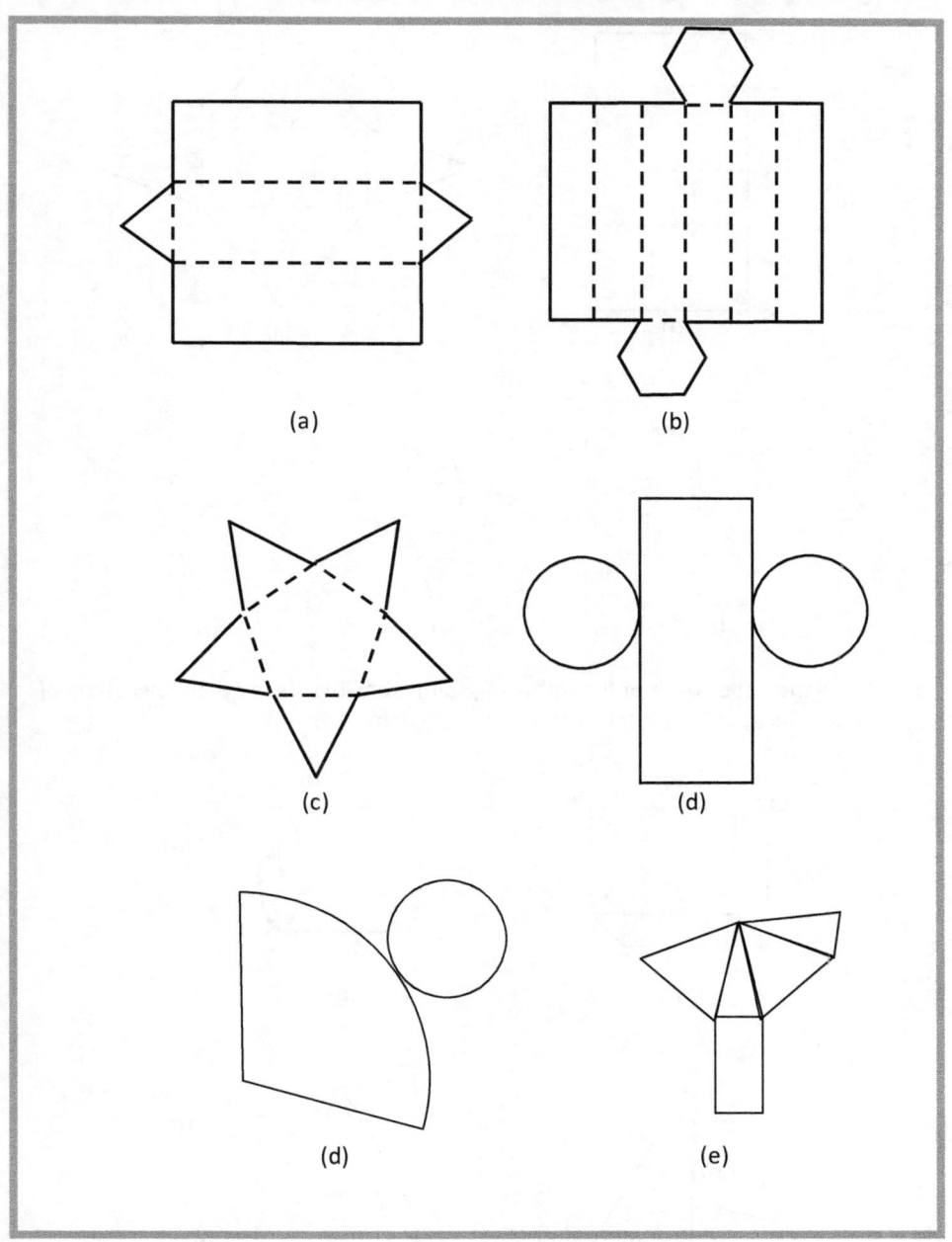

(a)

(b)

(c)

(d)

(d)

(e)

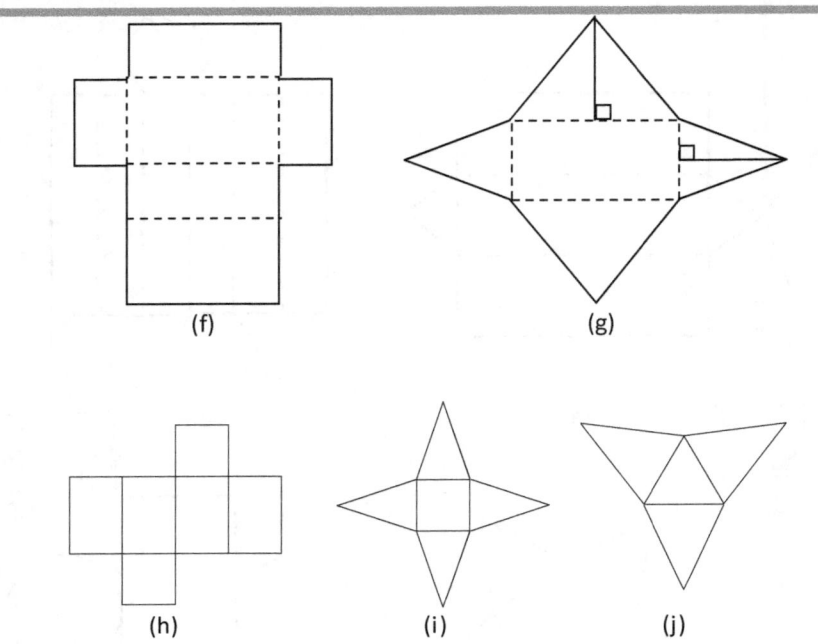

(f)   (g)

(h)   (i)   (j)

(5) Draw the net of each of the following solids showing clearly the dimensions of each. Hence, calculate the surface area of each solid.

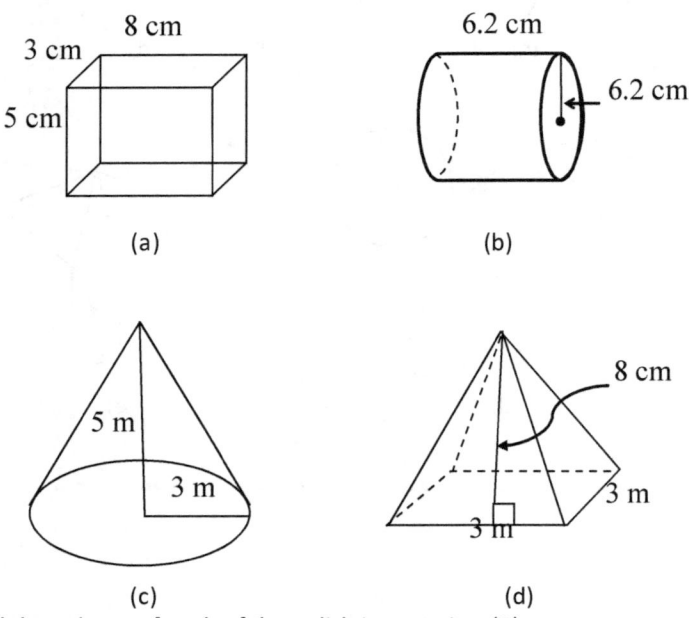

(a)   (b)

(c)   (d)

(6) Find the volume of each of the solids in question (5).

(7) Find the surface area and volume of a ball whose radius is 10.5 cm.

5.2 Surface Areas and Volumes of Similar Solid Figures

Let the cylinder C_2 be an enlargement of C_1, with scale factor k.

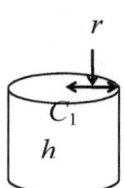 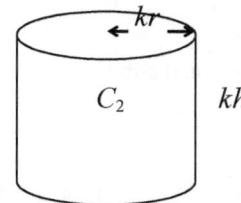

Surface area of $C_1 = \pi r^2 + 2\pi rh$ ... ... .①
Surface area of $C_2 = \pi(kr)^2 + 2\pi(kr)(kh)$
$\Rightarrow$ Surface area of $C_2 = (\pi r^2 + 2\pi rh)k^2$ ... ... ②
Dividing equation ② by equation ①

$$\frac{\text{Surface area of } C_2}{\text{Surface area of } C_1} = k^2$$

Volume of $C_1 = \pi r^2 h$ ... ... .③
Volume of $C_2 = \pi(kr)^2(kh)$ ... ... ④
Dividing equation ④ by equation ③

$$\frac{\text{Volume of } C_2}{\text{Volume of } C_1} = k^3$$

Therefore if the ratio of corresponding sides of two similar solid figures is $1:k$, then

(a) The ratio of their surface areas is $1:k^2$.

(b) The ratio of their volumes is $1:k^3$.

 Example

1. A cube, whose volume is $20\,\text{cm}^3$, is enlarged such that its volume now is $V\,\text{cm}^3$. Given that, the scale factor of the enlargement is 4. Find the value of V.

Solution

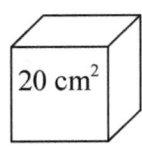

20 cm^2

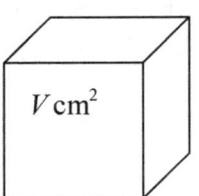

$V \text{ cm}^2$

Let V_0 = volume of original cube

Then, $\dfrac{V}{V_0} = k^3 \Rightarrow V = k^3 V_0$

$k = 4$ and $V_0 = 20 \text{ cm} \Rightarrow V = 4^3 \times 20 = 1280 \text{ cm}^3$

2. The ratio of the corresponding sides of a water tank and its model is $50:1$. Calculate the surface area of the tank in square metres given that the area of the model is 400 cm^2.

    Solution

    $\dfrac{\text{Area of tank}}{\text{Area of model}} = \dfrac{50^2}{1^2}$

    $\Rightarrow$ Area of tank $= 50^2 \times$ Area of model $= 50^2 \times 400 = 1000000 \text{ cm}^2$

    $\Rightarrow$ Area of tank $= 100 \text{ m}^2$.

5.3   Scale Drawing of Solid Figures

To draw to scale an enlargement of a figure, multiply corresponding sides by the scale factor. For instance, if the scale factor is k and one side is n units then the corresponding enlarged side will be nk. On the contrary, if the scale drawing is a reduction, divide corresponding sides by the scale factor. For instance, if the scale factor is k and one side is n units then the corresponding reduced side will be $\dfrac{n}{k}$.

Example

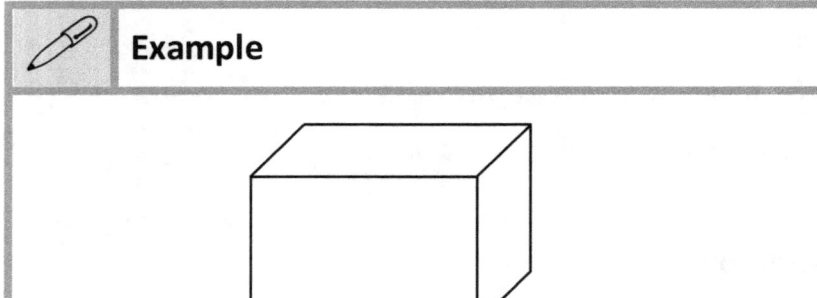

The diagram shows a cuboid.
(a) Draw a cuboid which is eight times by volume. Explain your reasoning.

(b) Your quarter has to build a similar water tank which can contain 375 m³ of water. They have local bricklayers who can build the tank but cannot determine the dimensions of the tank. Help them to work out the dimensions of the tank.

Solution

(a) Let volume of cuboid $= V_1$
And volume of scale drawing $= V_2$ $\Biggr\} \Rightarrow \dfrac{V_2}{V_1} = \dfrac{8}{1}$

Let the scale factor be k, then $\dfrac{V_2}{V_1} = k^3 \Rightarrow k^3 = 8$ and $k = 2$

By measuring the sides, we obtain

Length $= 3$ cm $\Rightarrow$ Length of scale drawing $= 2 \times 3$ cm $= 6$ cm
Height $= 1$ cm $\Rightarrow$ Height of scale drawing $= 2 \times 2$ cm $= 4$ cm
Width $= 2$ cm $\Rightarrow$ Width of scale drawing $= 2 \times 1$ cm $= 2$ cm

Therefore, the required cuboid shown below has length 6 cm, width 2 cm and height 4 cm.

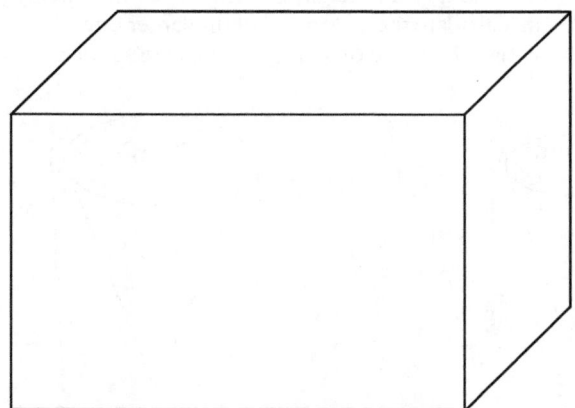

Volume of tank length $\times$ width $\times$ height $= (3k)(2k)(k) = 6000000$

$$\Rightarrow 6k^3 = 6000000 \Rightarrow k = \sqrt[3]{1000000} = 100$$

length $= 3 \times 100 = 300$ cm $= 3$ m
width $= 2 \times 100 = 300$ cm $= 2$ m
height $= 1 \times 100 = 300$ cm $= 1$ m

Exercise 4:4

1. Two similar containers have heights of 3 cm and 5 cm respectively. If the capacity of the smaller container is 54 cm^3, find the capacity of the larger container.

2. Two similar flasks are such that the ratio of the volume of the larger to the smaller is 7:2. Calculate the volume of the larger flask if that of the smaller is 56 cm^3. Given that the height of the smaller is 10 cm, find the height of the larger.

3. The following figure, shows two similar bowls with their volumes and the diameter of the smaller one given. Calculate the diameter d cm of the larger bowl.

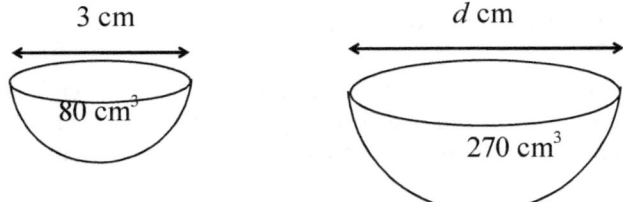

4. Two similar buckets have shapes in the form of a frustum. The volume of the smaller one is 24 cm^3 and its slant height is 6 cm. Given that the larger one has a slant height of 9 cm, calculate the volume V of the larger one.

5. In the figure below, find the value of V, if the cones are similar.

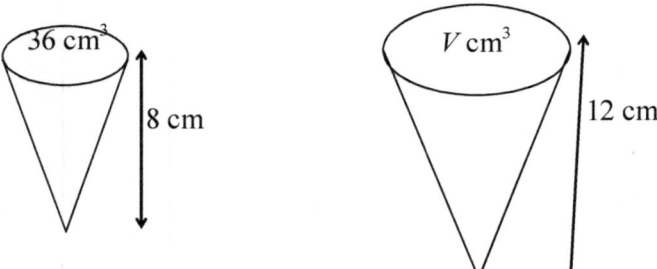

6. Given that the following cylinders are similar, find the value of y to the nearest tenth.

7. If the ratio of the diameters of two similar cups is 4:25, find the ratio of their volumes.

8. The surface area of a container of capacity 12.8 liters is 5000 cm^2. Find the

surface area of a similar contain with a capacity of 5.5 liters.

9. Two similar cylindrical watering cans have diameters of 35 cm and 45 cm. Find the volume of the larger watering can if the volume of the smaller watering can is 14450 cm³.

10. An engineer is drawing plans for a water tower. The tower is 41 m tall and the tank is circular with a diameter of 13 m.
    (a) The engineer builds a model of the tower with a scale of 1 cm: 5 m. What are the dimensions of the model?
    (b) You are employed to build a second model with height 10 cm. What is the scale for the model?

11. Njong wants to make a scale model of an elephant 7 m long, 3 m high with teeth of length 15 cm. The model should be not more than 20 cm tall. What scale should he use? Explain.

 # Multiple Choice Exercise 5

In this exercises, where necessary, take $\pi = \frac{22}{7}$.

1. The shape of each side of a cuboid is:
   [A] A triangle   [B] A trapezium   [C] A circle   [D] A rectangle

2. The number of vertices in a cuboid is:
   [A] 4   [B] 6   [C] 8   [D] 12

3. The number of faces in a cuboid is:
   [A] 4   [B] 6   [C] 8   [D] 12

4. The number of edges in a cuboid is:
   [A] 12   [B] 8   [C] 6   [D] 4

5. The total surface area of a cube of edge 3 cm is:
   [A] 27 cm² [B] 27 cm³ [C] 54 cm² [D] 36 cm²

6. The sides of two cubes are in the ratio 2:5. The ratio of their volumes is:
   [A] 4:5 [B] 8:15 [C] 6:125 [D] 8:125

7. A rectangular tank 2.25 m long and 1.6 m wide contains 2800 litres of water. Correct to the nearest cm, the depth of water in the tank is:
   [A] 76 cm [B] 78 cm [C] 770 cm [D] 780 cm

8. A cylindrical container closed at both ends, has a radius of 7 cm and a height 5 cm. The total surface area of the container is:
   [A] 154 cm² [B] 220 cm² [C] 528 cm² [D] 770 cm²

9. A cylindrical container closed at both ends, has a radius of 7 cm and a height 5 cm. The volume of the container is:
   [A] 154 cm³ [B] 220 cm³ [C] 528 cm³ [D] 770 cm³

10. The curved surface area of a cylindrical tin is 704 cm². The height when the radius is 8 cm is:
    [A] 3.5 cm   [B] 7 cm   [C] 14 cm   [D] 6 cm

11. Correct to 1 decimal place the volume of a cylinder of height 8 cm and base radius 3 cm is:
    [A] 300.0 cm³   [B] 250.0 cm³   [C] 226.2 cm³   [D] 150.9 cm³

12. The following figure shows a rectangular sheet of thin metal from which a cylinder, 10 cm high, is to be made with no overlap. The radius of this cylinder is:

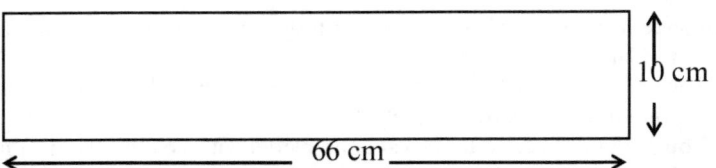

66 cm

10 cm

    [A] 3.3 cm   [B] 6.6 cm   [C] 10.5 cm   [D] 21 cm

13. A solid cylinder of radius 7 cm is 10 cm long. Its total surface area is:
    [A] 70π cm^2 [B] 18π cm^2
    [C] 210π cm^2 [D] 238π cm^2

14. The volume of a cylinder of radius 14 cm is 210 cm^3. The curved surface area of the cylinder is:
    [A] 30 cm^2 [B] 15 cm^2 [C] 616 cm^2 [D] 1262 cm^2

15. The internal and external radii of a cylindrical bronze pipe are 1.5 cm and 2 cm respectively. If the pipe is 10 cm long, the volume of the bronze used is:
    [A] $5\frac{1}{2}$ cm^3 [B] 55 cm^3 [C] $196\frac{2}{5}$ cm^3 [D] 550 cm^3

16. The cross-section of a prism is a right angled triangle 3 cm by 4 cm by 5 cm. The height of the prism is 8 cm. Its volume is:
    [A] 48 cm^3 [B] 60 cm^3
    [C] 96 cm^3 [D] 120 cm^3

17. The following figure shows a triangular prism of length 7cm. The right angled triangle PQR is a cross section of the prism $PR = 5$ cm and $RQ = 3$ cm. The area of the cross-section is:
    [A] 4 cm^2 [B] 6 cm^2 [C] 15 cm^2 [D] 20 cm^2

7 cm R

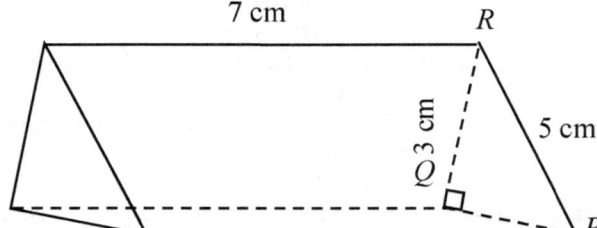

3 cm   5 cm

Q P

18. The figure above shows a triangular prism of length 7cm. The right angled triangle PQR is a cross section of the prism $PR = 5$ cm and $RQ = 3$ cm. The volume of the prism is:
    [A] 28 cm^3 [B] 42 cm^3 [C] 70 cm^3 [D] 84 cm^3

19. A prism is solid figure with:

    [A] regular faces   [B] uniform cross-sectional area.
    [C] triangular faces   [D] a square base and regular triangular faces.

20. The figure that is certainly not a prism is:

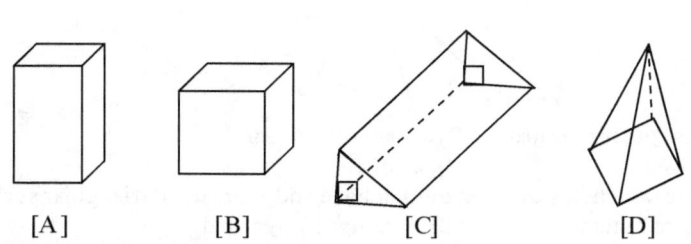

[A]   [B]   [C]   [D]

21. A solid with two parallel and congruent bases *cannot* be:
    [A] cone [B] prism [C] cylinder [D] cube

22. The solid formed by the net below is:
    [A] hexagonal prism   [B] rectangular pyramid
    [C] hexagonal pyramid   [D] rectangular prism

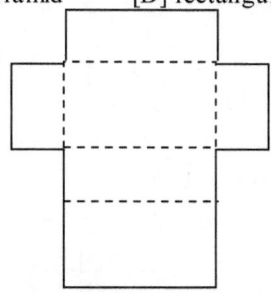

23. The height of a pyramid on a square base is 15 cm. Given that the volume is 80 cm^3, The length of the side of the base in cm is:
    [A] 3.3   [B] 5.3   [C] 4.0   [D] 8.0

24. The height of a pyramid on a square base is 15 cm. If the volume is 80 cm³, the area of the square base is:
    [A] 16 cm^2 [B] 9.6 cm^2 [C] 8 cm^2 [D] 25 cm^2

25. A right pyramid is on a square base of side 4 cm. The height of the pyramid is 3 cm. The volume of the pyramid is:
    [A] 9   [B] 4   [C] 8   [D] 16

26. A pyramid on a square base of side 10 cm has a height of 15 cm, its volume must be:
    [A] 150 cm^3 [B] 500 cm^3 [C] 1500 cm^3 [D] 5000 cm^3

27. The base of a pyramid is a 12 cm by 12 cm. If its height is 20 cm, the volume of the pyramid in cm^3 is:
    [A] 960   [B] 80   [C] 1440   [D] 1600

28. The net below is a net of:
    [A] a tetrahedron [B] a pyramid [C] a cone [D] a triangular prism

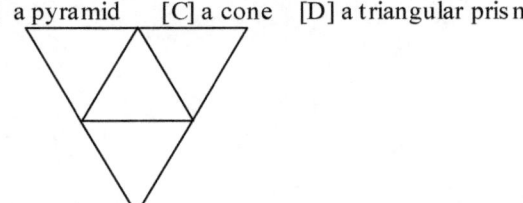

29. The following figure is called:

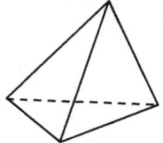

[A] a triangular pyramid   [B] a triangular prism
[C] a rhombus   [D] a cone

30. The figure which has one rectangular base and four lateral triangular surfaces is:
   [A] square pyramid   [B] rectangular pyramid
   [C] cone   [D] rectangular prism

31. Neither figure (a) below nor the following diagrams are drawn to scale.

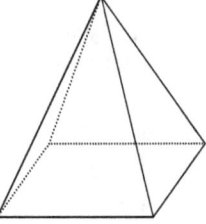

The net which corresponds to the figure above is:

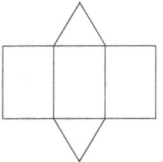   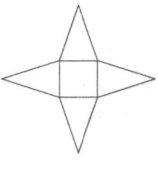   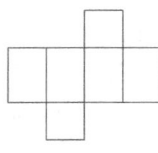   

    [A]   [B]   [C]   [D]

Answers to Structural Exercises

Exercise 1:1

1. (a) $(-2, -2)$ (b) $\left(-\dfrac{15}{2}, -4\right)$

2. (a) $\left(x, \dfrac{y_2 + y_1}{2}\right)$ (b) $\left(\dfrac{x_2 + x_1}{2}, y\right)$

3. (a) $(8, 1)$ (b) $\left(\dfrac{1}{2}, \dfrac{1}{2}\right)$ (c) $\left(-\dfrac{7}{2}, -\dfrac{7}{2}\right)$

    (d) $\left(\dfrac{3}{4}, -\dfrac{1}{2}\right)$ (e) $\left(\dfrac{3\sqrt{2}}{2}, 2\sqrt{3}\right)$ 4. 13 units

Exercise 1:2

1. $(1, 5)$ 2. $\left(6, \dfrac{19}{2}\right)$ 3. $(5, 1)$ 4. $(-16, 20)$

Exercise 1:3

1. (a) $\sqrt{2}$ (b) 13 (c) 5 (d) $\sqrt{205}$ (e) 2

2. $AB = 13$, $AC = 17$, $BC = 5\sqrt{2}$ 3. $-3 \pm 3\sqrt{3}$ 4. $2x + y = 14$

5. $a(t^2 + 1)$ 6. $x^2 + y^2 - 6x - 8y + 9 = 0$ 7. $k = -17$ 8. $\dfrac{11}{2}$

Exercise 1:4

1. (a) $\dfrac{3}{7}$ (b) ∞ (c) 1 (d) $\dfrac{3}{4}$ (e) $\dfrac{8}{15}$ (f) $\dfrac{3}{4}$ 2. $k = 11$ 3. $k = \pm 4$

4. (a) The points lie on a straight line

    (b) At $(0, 1)$ (c) $\left(-\dfrac{1}{2}, 0\right)$ (d) (i) 2 (ii) 2 (iii) 2

    (e) The gradients are the same. This suggests that we can calculate the gradient of a line using any two points on the line.

5. (a) $-\dfrac{3}{2}$ (b) $\dfrac{5}{6}$ (c) $\dfrac{4}{3}$ (d) $-\dfrac{4}{9}$

Exercise 1: 5

1. (a) $y = -\dfrac{2}{3}x + \dfrac{11}{3}$ (b) $13y = 15x - 1$ (c) $4y = 26x + 29$

   (d) $5y - x = 33$ (e) $y = x + \sqrt{2}$

2. $y = -\dfrac{1}{3}x + 1$ 3. $5y = -2x + 11$ 4. (a) $3y = x - 5$ (b) $y = -3x + 5$

5. (a) $7x = -3x + 1$ (b) $3y = 7x + 17$ 6. $5y = -12x + 29$

7. (a) $(2,3)$ (b) $8:5$ (c) $3y = 5x - 33$ 8. $y = -4x + 11$ 9. $b = \pm 4$

10. $n^2 = 9m^2$ 11. $an = mb$

12.   (a) A and B (b) A and C (c) None

Exercise 1:6

2. (c) $-1 < x < 5$

3. (i) ± 1 (ii) $x = \pm 1.5$ (iii) $x = 0$ or $x = 4$ (iv) $x = 0$ or $x = -3.5$

   (v) $x = -2$ or $x = 4$ (vi) $x = 1.5$

4. (i) $x = -8, x = 1$ (ii) $x = -\dfrac{3}{4}, x = 1$ (iii) $x = -7.4, x = 0.4$

   (iv) $x = -5.7, x = 0.7$ (v) $x = -1, x = -0.6$ (vi) $x = -4.4, x = 0.4$

Exercise 1:7

1.(a)

x	-5	-4	-3	-2	-1	0	1	2	3
y	5	1.5	-1	-2.5	-3	-2.5	-1	1.5	5

   (c) $(-1,-3)$ (d) -4.1 or 2.1 (f) $f(x) = x^2 - 5 - \dfrac{1}{x}$

2.(a)

x	0.25	0.5	1.0	1.5	2	2.5
y	4.5	3	3	3.7	4.5	5.4

   (c) 2.8 (d) -2 (e) 0.35 or 1.39

3. (a) 1.2 (b) $-2.4, -16.5$ (d) $-0.4, 2.7$ (e) 1.2

4.(a)

x	-2	-1	0	1	2	3	4	5
y	21	4	-7	-12	-11	-4	9	28

   (c) $-0.7, 3.4$ (d) -1.9 or 4.6 (e) 10

5.(a)

x	-2	-1	0	1	2	3	4	5
$f(x)$	20	7	-2	-7	-8	-5	2	13

(b) (i) -0.3 or 3.8 (ii) $f(x)_{min} = -8.1, x = 1.8$ (iii) $-0.3 < x < 3.8$ (c) -2.9 or 0.1

Exercise 1:8

1. 1

2.(a)

x	-4	-3	-2	-1	0	1	2	3
y		13		-1	-2	1	8	19

(c) 11

3.(a)

x	−1	0	1	2	3	4	5	6
y	10	−1	−8	−11	−10	−5	4	17

(c) 3   4. (b) −4

Exercise 1:9

1. (2,9), minimum

2. (a) −2   (b) $-\frac{1}{3}$ and 2   (c) $\left(\frac{5}{6}, -\frac{49}{12}\right)$, minimum point.

Exercise 1:10

1. 7.2 km/h,   77 m

2.

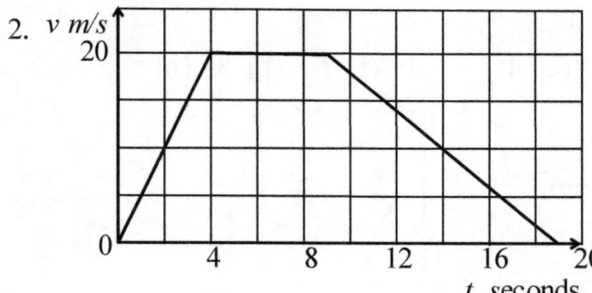

Distance covered = 240 m

3.

Final acceleration = 3 m/s²,
Final displacement = 114 m

4. (a) 11:20 a.m.   (b) 11:50 a.m.   (c) 12:20 p.m.   (e) 5.3 km

Exercise 2: 1

(a) $x = 70°$

(b) $x = 72°$, $y = 72°$

(c) $x = 80°$, $y = 100°$

(d) $x = 70°$, $y = 50°$

Exercise 2: 2

(1) 89°   (2) 40°   (3) 6   (4) 80   (5) 12   (6) 12

(7) 144°   (8) 12   (9) (a) Yes, 18   (b) No   (c) Yes, 24.

(10) 10   (11) 108°   (12) $\angle AFC = 45°$ (13) 12, $\angle ACD = 135°$

(14) (a) $\angle ABC = 140°$, $\angle CAE = 40°$, $\angle ACG = 60°$ (b) 100°

(15) $\angle PQR = 135°$, $\angle RSP = 45°$ (16) 60°

Exercise 2: 3

1. (a) 6   (b) 5   (c) 8

2. See next column

3. (i) (a) b, g, h, I   (b) b, d, e, g, h

   (ii) (a) N   (b) 2   (c) N   (d) 1   (e) 1   (f) N   (g) 8   (h) 4   (i) N

4. (i)

(a)   (b)   (c)

(ii)

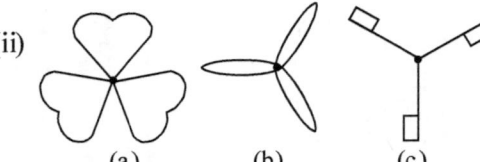

(a)   (b)   (c)

(iii)

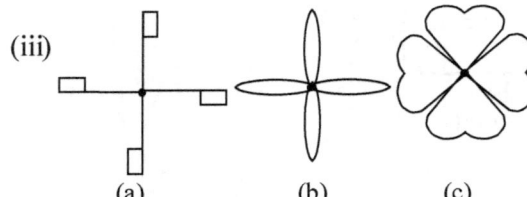

(a)   (b)   (c)

2.

Name of figure	Diagram	No. of lines of symmetry
(a) Kite		1
(b) Equilateral triangle		3
(c) Square		4
(d) Rectangle		2
(e) Regular pentagon		5
(f) Rhombus		2
(g) Parallelogram		0
(h) Regular hexagon		6

5. A, B, C, D, E, I, K, M, T, U, V, W, Y.
6. (a) 9 (b) 8 (c) 8 (d) 4 7. H, O, X
8. F, G, J, L, N, P, Q, R, S, Z.

9. (a) (b) (c)

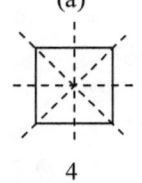 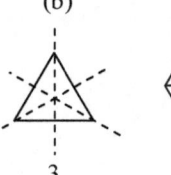 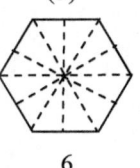

   4   3   6

(d)   (e)   (f)

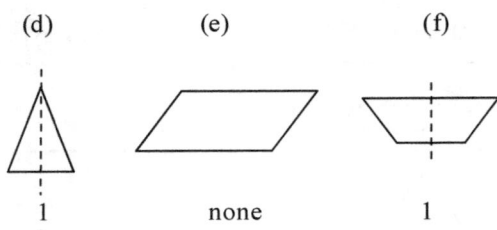

1   none   1

10. $H \cap V = \{$ H, I, O, X$\}$

    $H \cap V = R$ and all Members of $H \cap V$ have at least two lines of symmetry.

Exercise 3: 1

1. 13   2. 1   3. 9   4. (a) 54 kg (b) 51.2 kg (c) 54 kg
5. (a) 2.92   (b) 3   (c) 3   6. (a) 5.3   (b) 5   (c) 5
7. (a) 7   (b) 7.1 (c) 7   8. (a) 113.4 FRS (b) 100 FRS   (c) 50 FRS
9. (a) 70 kg   (b) 68 kg   (c) 70.25 kg   10. (a) 30   (b) 8   (c) 6.1
11. (a) 2   (b) 2   (c) 2   12. (a) 5   (b) 12   (c) 7.7

Exercise 3: 2

1. Mode. Stock goods are most needed.
2. Mean. It takes into account all the values.
3. Mean. It takes into account all the values.
4. The average is a misleading statistic because so many farmers have no pigs and one farmer has so many.

Exercise 3: 3

1. (a)

x	66	67	68	69	70	71
f	2	3	7	4	3	1

   (b) 68   (c)   68   (d)   68   (e) 68.3

2. (a)

GRADE	A	B	C	D	E	F
Angle of Sector	30	60	90	40	70	70

(b)

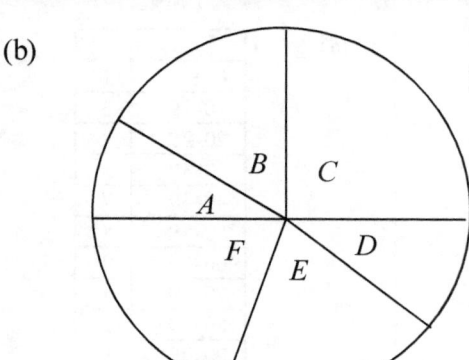

(c) (i) 72   (ii) 8   (d) 1:1

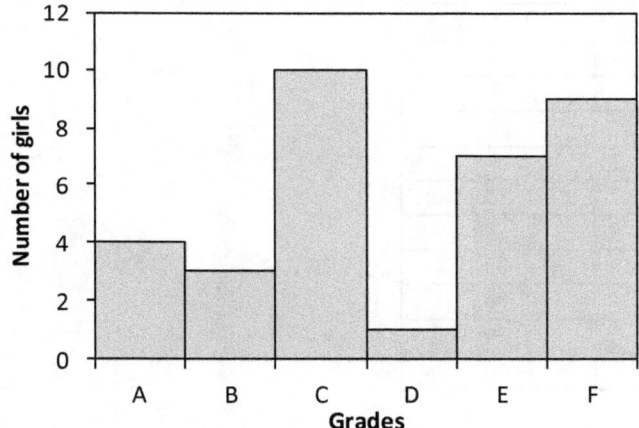

3. (i) (a)

Marks	Cum freq
≤ 10	0
≤ 20	2
≤ 30	8
≤ 40	15
≤ 50	29
≤ 60	49
≤ 70	84
≤ 80	113
≤ 90	119
≤ 100	120

4. (a)

x	4	5	6	7	8	9	10
f	9	3	7	4	5	1	1

    (b) 4   (c) 6

5. (a)

x	f
0- 9	1
10-19	2
20-29	2
30-39	3
40-49	7
50-59	8
60-69	9
70-79	4
80-89	3
90-99	1

(b)

x	f	Cum freq
≤ 10	1	1
≤ 20	2	3
≤ 30	2	5
≤ 40	3	8
≤ 50	7	15
≤ 60	8	23
≤ 70	9	32
≤ 80	4	36
≤ 90	3	39
≤ 100	1	40

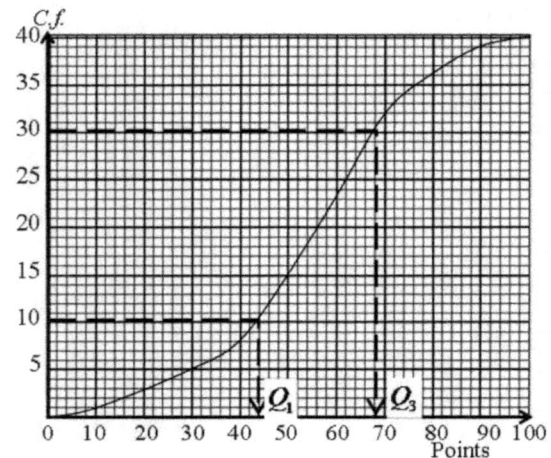

(c) 60-69   (d) 54.2   (e) 24   (f) A grade

6. (a)

x	f
35-44	3
45-54	7
55-64	13
65-74	16
75-84	7
85-94	4

(b)

x	f	
< 45	3	3
< 55	7	10
< 65	13	23
< 75	16	39
< 85	7	46
< 95	4	50

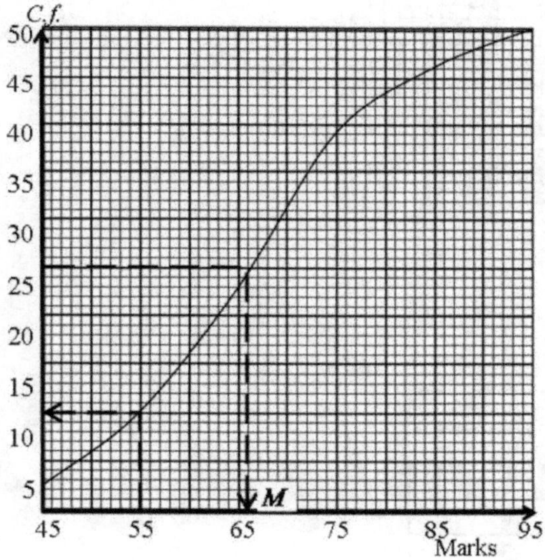

(c) 66   (d) 80%

7. (a) 43.6,   (b) 86.4   (c) 26   (d) −10

8. (i) 34.5 minutes

(ii)

Time (min)	Cum freq
$t \le 20$	0
$t \le 25$	8
$t \le 30$	16
$t \le 35$	28
$t \le 40$	58
$t \le 45$	76
$t \le 50$	80

(ii)

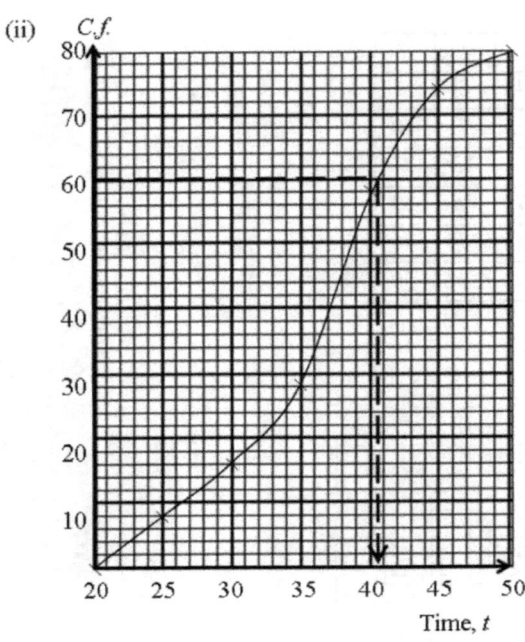

C.f.

(iii) 40.5 minutes

9.   41.1, 11.9

10.

No. of times	No. of students
0 - 5	34
6- 10	4
11-15	10
16-20	25
21-25	50
26-30	37
31-35	24
36-40	15
41-45	1

(a) 21.5   (b)   113.7   (c)   10.7

11. (a) 45.79   (b) 21.18   12. 166.62, 166.55, 2.21,

13. (a) 44.5   (b)   240   (c)   15.5

14. (a) 34.4 years   (b) 32 years   (c) 17 years   (d) 186.5   (e) 13.7 years

15.(a) 44.58   (b)

No. of times	Cum frequency
< 29.5	40
< 39.5	160
< 49.5	360
< 59.5	460
< 69.5	480
< 79.5	496
< 89.5	500

(c)

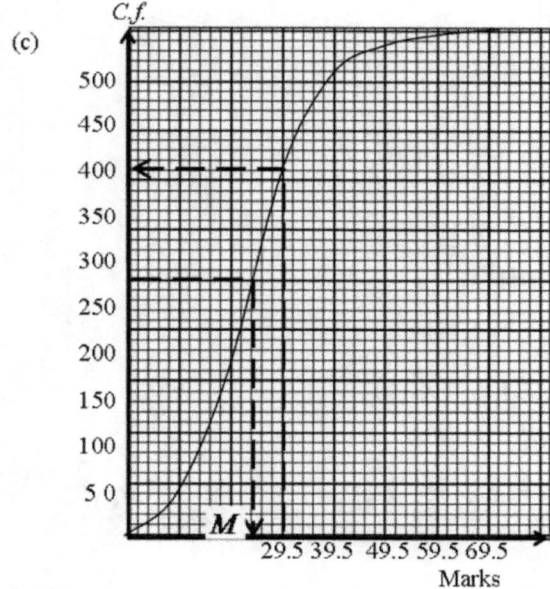

(i) 43.6   (ii)   360

Exercise 4: 1

1. (a) $\dfrac{1}{6}$ (b) $\dfrac{1}{2}$ (c) $\dfrac{1}{2}$ (d) $\dfrac{2}{3}$ (e) $\dfrac{1}{2}$

2. (a) $\dfrac{3}{8}$ (b) $\dfrac{1}{8}$ (c) 0 (d) $\dfrac{1}{8}$

3. $\dfrac{1}{4}$ 4. $\dfrac{3}{10}$ 5. $\dfrac{2}{11}$ 6. $\dfrac{1}{3}$

Exercise 4: 2

1. (a) $\frac{9}{35}$ (b) $\frac{23}{35}$ (c) $\frac{2}{5}$ 2. $\frac{11}{15}$ 3. $\frac{21}{25}$ 4. $\frac{3}{10}$ 5. $\frac{3}{7}$ 6. $\frac{1}{4}$

Exercise 4: 4

1. $\frac{2}{3}$ 2. $\frac{2}{3}$ 3. $\frac{2}{9}$ 4. $\frac{11}{18}$ 5. (i) $\frac{1}{3}$ (ii) $\frac{5}{18}$ (iii) $\frac{11}{18}$ 6. $\frac{17}{30}$

7. $\frac{2}{3}$ 8. (a) $\frac{2}{7}$ (b) $\frac{3}{7}$ 9. (a) $\frac{1}{13}$ (b) $\frac{4}{13}$ 10. (a) $\frac{2}{5}$ (b) $\frac{3}{10}$ (c) $\frac{3}{5}$

11. (a) $\frac{1}{6}$ (b) $\frac{1}{6}$ (c) $\frac{1}{3}$ 12. $\frac{1}{3}$ 13. 1   14. 1   15. $\frac{13}{15}$

Exercise 4: 5

1.(a) $\dfrac{9}{25}$ (b) $\dfrac{4}{25}$ (c) $\dfrac{12}{25}$ 2. $\dfrac{1}{36}$ 3.(a) $\dfrac{17}{35}$ (b) $\dfrac{18}{35}$

4.(a) $\dfrac{2}{11}$ (b) $\dfrac{3}{11}$ (c) $\dfrac{5}{11}$ (d) $\dfrac{6}{11}$

5.(a) $\dfrac{3}{8}$ (b) $\dfrac{11}{16}$ (c) $\dfrac{5}{16}$ (d) $\dfrac{1}{8}$ 6.(a) $\dfrac{1}{9}$ (b) $\dfrac{4}{9}$ (c) $\dfrac{4}{9}$

7.(a) $\dfrac{6}{25}$ (b) $\dfrac{9}{25}$ (c) $\dfrac{4}{25}$ (d) $\dfrac{6}{25}$ (e) $\dfrac{13}{25}$ 8.(a) $\dfrac{1}{12}$ (b) $\dfrac{7}{9}$ (c) $\dfrac{5}{18}$

9.(a) $\dfrac{8}{35}$ (b) $\dfrac{32}{105}$ (c) $\dfrac{79}{105}$ 10. $\dfrac{3}{4}$ 11. $\dfrac{5}{18}$ 12.(a) $\dfrac{9}{49}$ (b) $\dfrac{24}{49}$ (c) $\dfrac{144}{343}$

13. $\dfrac{11}{21}$ 14.(a) $\dfrac{9}{25}$ (b) $\dfrac{13}{25}$ 15. $\dfrac{5}{36}$ 16.(a) $24-m$ (b) $\dfrac{m+6}{30}$

17. 18   18.(a) $\dfrac{1}{26}$ (b) $\dfrac{6}{13}$ 19. 1   20. 1

Exercise 4: 6

1. (a) 10   (b) 30-39   (c) 47.7 kg

(d)

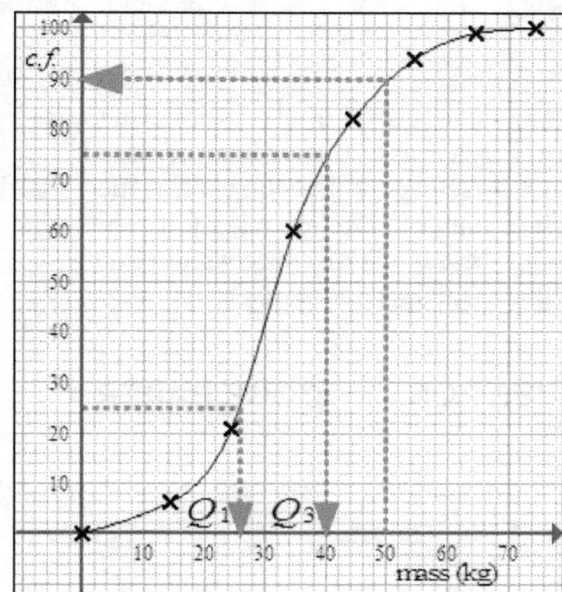

(e)   14 kg   (f) $\dfrac{9}{10}$

2. (a) 10 minutes   (b) 11 minutes   (c) 1.75   (d) $\dfrac{1}{5}$ (e) $\dfrac{5}{12}$

3. (i)   120-149   (ii) 135.7 kg

(iii)

Weights(g)	C.F.
<30	20
<60	45
<90	95
<120	170
<150	290
<180	390
<210	470
<240	500

(iii) See graph on the next page

(a) 140 g   (b) 70   (c) $\dfrac{6}{25}$

Exercise 5

1. $250\,cm^3$ 2. $268\,cm^3$, 35 cm   3. 4.5   4. $81\,cm^3$ 5. $121.5\,cm^3$

6. 10.5   7. 64:15625   8. $2847\,cm^3$ 9. $30712\,cm^3$

10. (a)   diameter = 2.6 cm and height = 8.2 cm   (b)   1 cm : 4.1 m

177

Examination Type Questions

The following are passed examination questions. Questions on topics which are not in the recent GCE syllabus 570 have been replaced.
Test yourself by going through these papers. In each case, respect the time given to work through the whole paper before referring to the answers. I bet you that after going through these questions successfully, no examination at this level will be an obstacle to you.

MULTIPLE CHOICE QUESTIONS 1 - CGCE PAPER 1 SAMPLE

Time allowed: One and a half $\left(1\frac{1}{2}\right)$ hours.

1. The value of $\dfrac{\left(5\times10^{-2}\right)\left(7\times10^{4}\right)}{2\times10^{5}}$ to two significant figures is:

   [A] 0.01   [B]   0.175   [C]   0.02   [D] 0.018

2. When simplified, the value of $1\frac{2}{5}+1\frac{3}{5}\div\frac{4}{5}$ is:

   [A] $3\frac{4}{5}$ [B] $2\frac{2}{5}$ [C] $3\frac{2}{5}$ [D] $2\frac{17}{25}$

3. The base of a cuboid of height 6cm is of length 12cm and its perimeter 32cm. The volume of the cuboid is:

   [A]   288 cm^{3} [B] 720 cm^{3} [C] 1440 cm^{3} [D] 2304 cm^{3}

4. The figure below represents the mapping from $\mathbb{R}$ to $\mathbb{R}$,
   $f : x \mapsto 2x+3$. The values of p and q are respectively:

   [A]   7 and 4   [B]   4 and 7   [C]   3 and 8   [D]   25 and 7

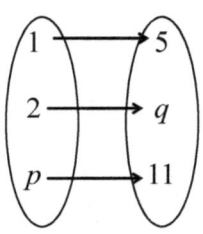

5. The number of sides of a regular polygon whose interior angles are each 140° is:

    [A] 40   [B] 9   [C] 5   [D] 3

6. Given that the lines l_1 and l_2 in figure 2 are parallel, the value of the angle marked a is:

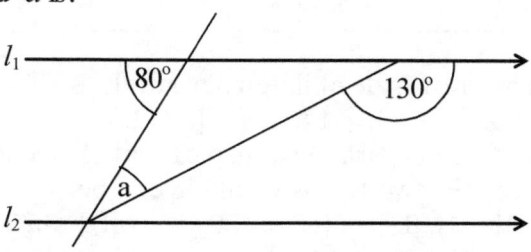

    [A] 60°   [B] 50°   [C] 30°   [D] 10°

7. A second hand car depreciates annually by 20% of its value at the beginning of the year. A car bought for 2.5 million francs in January 2003 will in December 2004 be valued at:

    [A] 2 million francs   [B] 1.7 million francs
    [C] 1.5 million francs   [D] 1.6 million francs

8. The sum of the first n terms of a sequence is given by $S_n = n(3n+5)$. The fifth term of the sequence is:

    [A] 32   [B] 38   [C] 68   [D] 100

9. The straight line $3x - 4y = 12$, cuts the axes at P and Q. The length of the line segment PQ is:

    [A] $\sqrt{7}$ [B] $\sqrt{5}$ [C] 5   [D] 12

10. In the figure below, the sum of the angles a, b and c is:

    [A] 225°   [B] 105°   [C] 285°   [D] 270°

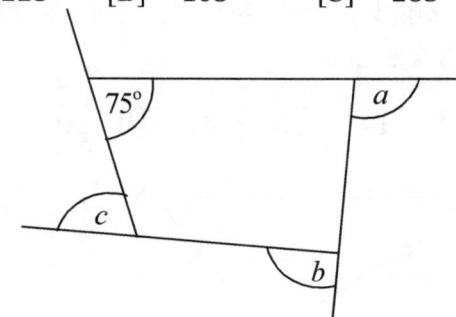

11. The vector $\mathbf{v}$ is given by $\frac{1}{2}(\mathbf{a} + 3\mathbf{b})$ where $\mathbf{a} = 4\mathbf{i} + 9\mathbf{j}$ and $\mathbf{b} = 2\mathbf{i} - 11\mathbf{j}$. The magnitude of $\mathbf{v}$ is:

    [A] $\sqrt{5}$ [B] 13   [C] 9   [D] 11

12. In a survey families were asked the number of male children they have and the results were as follows:

No. of male children	0	1	2	3
No. of families	4	8	6	2

The mean number of male children per family is:
[A] 1   [B] 2   [C] 1.6   [D] 1.3

13. Two towns are 64 km apart. On a map of scale 1:5000000, the distance between the two towns would in cm, be:
[A] 1.28   [B] 12.8   [C] 128   [D] 3.25

14. The figure below shows a table top made with two congruent trapezia. The area of the table top is:
[A] 550 cm² [B] 660 cm² [C] 696 cm² [D] 3300 cm²

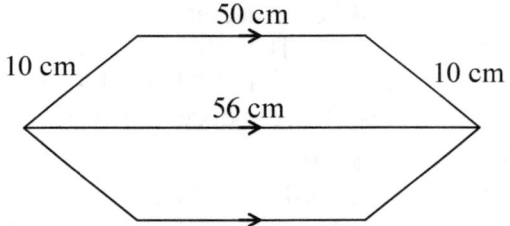

15. Given the sets P and R such that $P' \cap Q = \varnothing$, it follows that:
[A] $P \subset R$ [B] $P' = Q$ [C] $P \cup R = \mathscr{E}$ [D] $R \subset P$

16. *PQRS* as in Figure 5 is a parallelogram. Given that **PS = a, PQ = b** and $ST : TQ = 1 : 2$, vector **TR** equals:
[A] $\frac{1}{2}(\mathbf{a}+\mathbf{b})$ [B] $\frac{1}{3}(\mathbf{a}+\mathbf{b})$ [C] $\frac{1}{2}(3\mathbf{b}-\mathbf{a})$ [D] $\frac{1}{3}(4\mathbf{b}-\mathbf{a})$

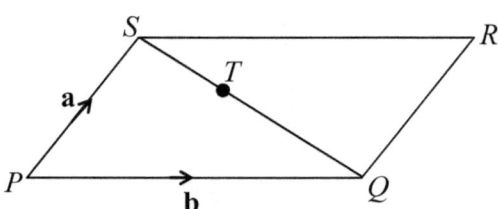

17. There four bottles of Top Orange and 6 bottles of Top Citron in a crate kept in a room. The bottles are all identical. In the dark, Bona goes in

and takes a drink. Cheryl follows and takes herds too. The probability that both drinks are of the same kind is

[A] $\frac{12}{90}$ [B] $\frac{42}{90}$ [C] $\frac{30}{90}$ [D] $\frac{12}{100}$

18. The shaded areas in the following figure represent:

[A] $(P \cap Q) + R$ [B] $P \cup Q \cap R$ [C] $(P \cap Q) \cup R$ [D] $P \cap Q \cap R$

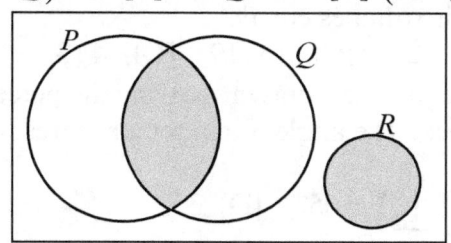

19. Three quantities F, G and H are such that $F:G = 3:4$ and $G:H = 1:2$. The ratio of F to H must be:

[A] $3:8$ [B] $8:3$ [C] $3:2$ [D] $6:4$

20. Given that $8^{(1+3x)} = 2^{(1-x)}$, the value of x is:

[A] $\frac{1}{7}$ [B] $-\frac{1}{7}$ [C] $\frac{1}{5}$ [D] $-\frac{1}{5}$

21. The point $(3,-4)$ lies on the line $y = 4x + c$. The equation of a line that passes through $(0, c)$ and is perpendicular to the first line is:

[A] $y = \frac{1}{4}x + 8$

[B] $y = 4x - 8$

[C] $y = -\frac{1}{4}x - 8$

[D] $y = -\frac{1}{4}x + 8$

22. In figure 7, CB is tangent to the circle with centre O. Given that $CB = 9$ cm and $AB = 6$ cm, the radius of the circle is:

[A] 6 cm [B] 7.5 cm [C] $\sqrt{4}$ cm [D] 3 cm

23. Given that $\tan \tan G = \frac{5}{12}$, the value of $\sin G + \frac{1}{2} \cos G$ is:

[A] $\frac{17}{13}$ [B] $\frac{18}{12}$ [C] $\frac{9}{12}$ [D] $\frac{11}{13}$

24. A coin is weighted such that the chance of a head showing on any toss is three times the chance of a tail showing. The probability of a tail showing on one toss is:

[A] $\frac{1}{2}$ [B] $\frac{1}{3}$ [C] $\frac{1}{4}$ [D] 0

25. A bicycle costing 60,000 FRS was bought at 10% discount. It was later sold to Mr. Brain and a profit of 20% was made. Mr. Brain paid:

[A] 78,000 FRS   [B] 72,000 FRS   [C] 66,000 FRS   [D] 64,800 FRS

26. The expression, when expressed as a single fraction is:

[A] $\dfrac{2x-1}{3x(2x-1)}$ [B] $\dfrac{4x-2}{6x-1}$ [C] $\dfrac{7x+1}{3x(2x-1)}$ [D] $\dfrac{8}{3(2x-1)}$

27. The surface areas of two similar solid cylinders are respectively 48 and 27 cm^2. The ratio of their volumes equals:

[A] $48:27$ [B] $64:27$ [C] $16:19$ [D] $4:3$

28. The following figure is a pie chart drawn showing the percentage sales of drinks in a certain week. The angle of the sector representing Amstel is:

[A] $126°$ [B] $159°$ [C] $165°$ [D] $35°$

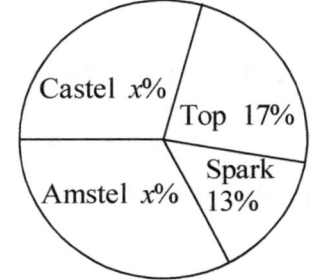

29. Three towers P, Q and R are located in a plain such that the bearing of P from Q is $332°$. Given that angle $PQR = 113°$ and R is due south of P, the baring of q from R is:

[A] $219°$ [B] $039°$ [C] $051°$ [D] $062°$

30. The position vector of the midpoint of the line segment PQ, where $P\,(5,2)$ and $Q\,(-1,-4)$, is equal to:

[A] $(2,-1)$ [B] $(4,-2)$ [C] $2\mathbf{i}-\mathbf{j}$ [D] $4\mathbf{i}-2\mathbf{j}$

31. Figure 9 is the day's record of sales of rice to retailers as recorded by the shopkeeper. The man number of bags bought pr retailer is:

[A] 35   [B] 25   [C] 30   [D] 15

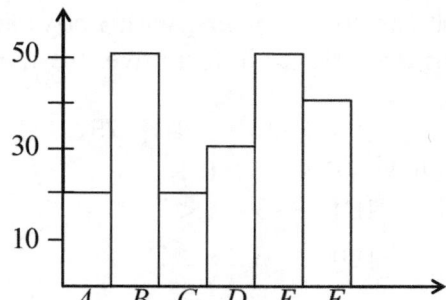

32. When converted to base ten, the value of $23\text{six} + 45\text{six}$ is:
    [A] 408   [B]   112   [C]   68   [D] 44

33. A triangle and a rectangle both have equal areas. The triangle has height 8 cm and base 6 cm. Given that the length of the rectangle is 10 cm, its with must be:
    [A] 12 cm   [B]   7.5 cm   [C]   4.8 cm   [D] 2.4 cm

34. Given that $p = \dfrac{3ab^2 - ab}{5}$, a, when expressed in terms of p and b, is equal to:

    [A] $\dfrac{5p}{3b^2 + b}$ [B] $\dfrac{5p}{b(3b-1)}$ [C] $\dfrac{5p}{5(3b^2 - b)}$ [D] $\dfrac{5p}{b(3b+1)}$

35. The function f and g are defined in $\mathbb{R}$ the set of real numbers as
    $$f : x \mapsto 4 - 3x$$
    $$g : x \mapsto 2x + 1$$
    It follows that $g^{-1} \circ f(x)$ equals:

    [A] $9 - 6x$ [B] $1 - 6x$ [C] $\dfrac{3}{2}(1 - x)$ [D] $\dfrac{1}{2}(14 - 3x)$

36. In the figure 10, $NK = 5$ cm, $MQ = 4$ cm and angles $NMQ = 60°$. Given that angles MQN and MPK are right angles, the length of MP is equal to:

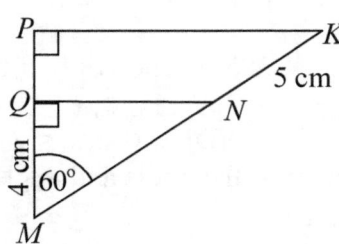

    [A] 6.5 cm   [B]   4.5 cm   [C]   7 cm   [D] 9 cm

37. The school science club has 300 members, with 3 boys for very 2 girls. Given that 25% of the girls are in form five, the number of form five girls in the club is:
    [A] 50   [B]   30   [C]   90   [D] 75

38. The range of x for which $x^2 - 2x - 15 \geq 0$ is:
    [A] $x \leq -3$ or $x \geq 5$ [B] $-3 < x < 5$

    [C] $x \leq -3$ or $x \leq 5$ [D] $-3 \leq x \leq 5$

39. The set M, K and P are the subsets of $\mathbb{N}$ the set of natural numbers, and defined as

    $M = \{x : x$ is a factor of 12$\}$

    $K = \{x : x$ is a prime factor of 12$\}$

    $P = \{x : x$ is an even number less than 12$\}$

    The Venn diagram illustrating the relationship between these sets is:

    [A] 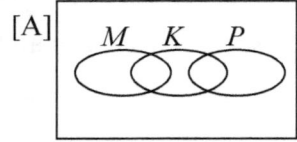   [B]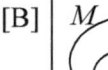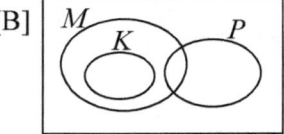

    [C] 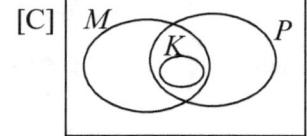   [D]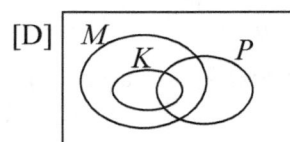

40. Sugar is sold for 650 francs a box or for a pack of 5 boxes. By buying 15 boxes at once, a woman saved the sum of:
    [A] 9,000 FRS   [B] 9750 FRS   [C] 750 FRS   [D] 1500 FRS

41. Given the equations $2m + 7 = n$ and $5n - 2m = 19$, the value of n that satisfies both equation is:
    [A] -2 [B] $6\frac{1}{2}$ [C] 3 [D] $3\frac{1}{2}$

42. The relation $g : x \mapsto \sqrt{x^2 - 4}$ is a function only if the domain is:
    [A] $x \in \mathbb{R}, x \geq 2$ [B] $x \in \mathbb{R}, x \geq 2, x \leq -2$
    [C] $x \in \mathbb{R}$ [D] $x \in \mathbb{R}, x \leq 2$

43.   In figure 11, O in the centre of the circle and angle $SRQ = 150°$. The value of the angle marked x is:

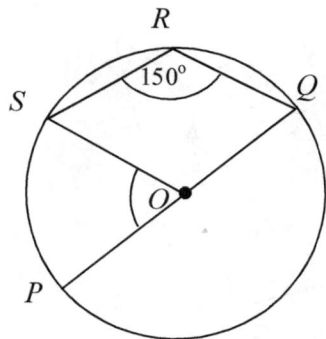

[A]   150°   [B]   120°   [C]   300°   [D]   60°

44. Given a circle of area 616 cm², and taking $\pi = \dfrac{22}{7}$, the circumference

of the circle must be:

[A] 196 cm   [B] 154 cm   [C] 44 cm   [D]   88 cm

45. The functions s and t are defined thus:
$$s: x \longmapsto 3x - 1$$
$$t: x \longmapsto k - 2x, k \in \mathbb{Z}$$

Given that $t(-2) = 7$, the image of 3 under the composite function $t \circ s$ is:

[A]   −10   [B]   −13   [C]   −6   [D]   14

46. The equation of the line that passes through the point $(2, -7)$ and is parallel to the line $3x + y = 5$ is:

[A] $y = -3x - 1$ [B] $y = 3x - 1$

[C] $3x + y = 4$ [D] $3y = x - 23$

47. The quantity p is known to vary inversely as the square of the quantity q. Given that p is 8 when q is 2, the value of q when p is 2 is:

[A]   8   [B]   4   [C]   1   [D] $\dfrac{1}{2}$

48. Triangle PQR in figure 12 is an enlargement of triangle KLM. Given that $KL = 5$ cm $PQ = 15$ cm and $QR = 13.5$ cm, the length of LM is:

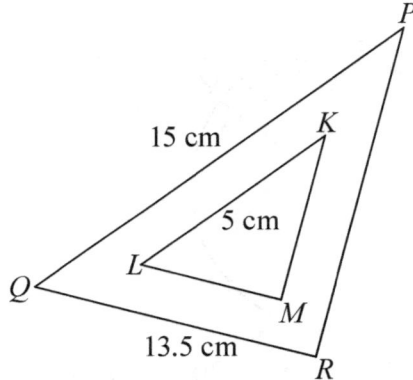

[A]   2.7 cm   [B]   3.5 cm   [C]   4.5 cm   [D]   1.5 cm

49. The quadratic curve C cuts the x-axis at the points $(-3,0)$ and $(2,0)$. Given that the curve has maximum point, its equation is:

[A] $y = x^2 + x - 6$ [B] $y = x^2 - 3x + 2$

[C] $y = x^2 + 3x - 2$ [D] $y = 6 - x - x^2$

50. Mole's house is 30 m due west of the public tap. While Bobe's house is due north of the same tap. The distances between the two houses are 50 m. The difference between the distances covered by the wives of both men to get to the tap is therefore:

[A]   10 m   [B]   20 m   [C]   30 m   [D]   40 m

MULTIPLE CHOICE QUESTIONS 2 - NWR MOCK 2015 PAPER 1

Time allowed: One and a half $\left(1\frac{1}{2}\right)$ hours.

1.   The value of $6 + 2 \times 4 - 15 \div 3$ is:
[A]   3   [B]   9   [C]   4   [D]   27

2.   the value of the digit 4 in the number 27.94 is:
[A]   4 unit   [B]   4 hundred [C]   4 tenth   [D]   4 hundredth

3.   In the network below, the number of vertices, regions and arcs respectively is:
[A]   4, 6, 8   [B]   6, 4, 8   [C]   8, 4, 6   [D]   8, 6, 4

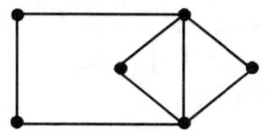

4. A equivalent fraction to $\frac{4}{7}$ is:

   [A] $\frac{16}{28}$ [B] $\frac{9}{12}$ [C] $\frac{24}{28}$ [D] $\frac{7}{4}$

5. Expressing 80 % as the product of its prime factors gives:

   [A] $2^3 \times 5$ [B] 2×5^3 [C] $2^2 \times 5^2$ [D] $2^4 \times 5$

6. The scale of a map is $1: 500,000$. The distance on the map between two points 25km apart is:

   [A]   5 cm   [B]   5 m   [C]   5mm   [D]   5dm

7. Given the formula $T = \sqrt{r^2 + 1}$

   [A] $r = \sqrt{T^2 - 1}$ [B] $r = \sqrt{T^2 + 1}$ [C] $r = \sqrt{T - 1}$ [D] $r = \sqrt{1 - T^2}$

8. The solution set of the quadratic equation $x^2 + 4x - 5 = 0$ is:

   [A] $\{1, -5\}$ [B] $\{-1, 5\}$ [C] $\{-1, -5\}$ [D] $\{1, 5\}$

9. A school scored 60 % in an examination. If 72 students passed then the number of students who sat for the examination is:

   [A]   60   [B]   100   [C]   72   [D]   120

10. Factorizing $4x^2 - 9$ gives:

    [A] $(2x - 3)(2x - 3)$ [B] $(2x + 3)(2x + 3)$
    [C] $(2x - 3)(2x + 3)$ [D] $x(4x - 9)$

11. If $\sin \theta = \frac{3}{5}$, then $\tan \theta$ is:

    [A] $\frac{4}{5}$ [B] $\frac{4}{3}$ [C] $\frac{5}{3}$ [D] $\frac{3}{4}$

12. One of the subsets of the set $= \{1,2,3\}$ is:

    [A] $\emptyset$ [B]   1   [C]   2   [D]   3

13. In the figure below, the shaded portion can be best described as:

    [A] $(A \cap B)'$ [B] $A \cup B$ [C] $A' \cup B'$ [D] $(A \cup B)'$

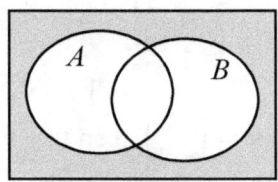

14. Given two statements p: Bih is in Bamenda and
    q: Bamenda is in Cameroon, then:

    [A] $q \Rightarrow p$ [B] $p \Rightarrow q$ [C] $\sim p \Rightarrow q$ [D] $\sim q \Rightarrow p$

15. Given that $A = \{1,2,3,4\}$ then $n[P(A)]$ is:

    [A] 8   [B] 4   [C] 16   [D] 12

16. Given that $\log_3 x = 2$, the value of x is:
    [A] 9   [B] 8   [C] 6   [D] 0

17. Given that, $3^x = \dfrac{1}{243}$ the value of x is:
    [A] 5   [B] $\dfrac{1}{5}$ [C] 3^{-5} [D] -5

18. The value of k for which $\begin{pmatrix} 5 & 6 \\ 1 & 4 \end{pmatrix} = \begin{pmatrix} 5 & 2k \\ 1 & 4 \end{pmatrix}$ is:
    [A] 6   [B] 5   [C] 3   [D] 7

19. The remwhen the expression $f(x)$ where $f(x) = x^2 + x - 5$ is divided by $(x - 2)$ is:
    [A] -5 [B] -3 [C] -1 [D] 1

20. Given the function f defined as $f: x \longmapsto 2x - 1$, the value of $f(-2)$ is:
    [A] -2 [B] -5 [C] 4   [D] 5

21. The domain D_f of the function $f: x \longmapsto \dfrac{5}{x-2}$ is:
    [A] $\mathbb{R}$ [B] $\mathbb{R} - \{-2\}$ [C] $\mathbb{R} - \{2\}$ [D] $\mathbb{R} - \{5\}$

22. Given the function $g: x \longmapsto 2x + 1$, then the value of $g^{-1}(7)$ is:
    [A] 15   [B] 3   [C] 4   [D] 12

23. Given the function h such that $h: x \longmapsto x + 3$, the composite function $hh^{-1}(x)$ is:
    [A] $(x + 3)(x - 3)$ [B] $(x + 3)(x + 3)$ [C] x [D] $x - 6$

24. The gradient of a line perpendicular to the line $y - 2x = 5$ is:
    [A] $-\dfrac{1}{2}$ [B] -5 [C] $\dfrac{1}{2}$ [D] 2

25. The length of the line segment AB with coordinate with coordinates $A(3,7)$ and $B(7,4)$ is:
    [A] $\sqrt{58}$ units   [B] 4 units   [C] $\sqrt{65}$ units   [D] 5 units

26. The coordinates of the midpoint of the line linking the points $P(3,7)$ and $T(7,3)$ is:
    [A] $(1,-3)$ [B] $(2,6)$ [C] $(6,-3)$ [D] $(-6,6)$

27. The gradient of the line which passes through the points $(6,3)$ and $(-2,19)$ is:
    [A] 4   [B] -2 [C] -4 [D] 2

28. The modulus of the vector $\mathbf{r} = \begin{pmatrix} -1 \\ 1 \end{pmatrix}$ is:
    [A] 0   [B] 2   [C] $\sqrt{2}$ [D] $\sqrt{-2}$

29. Given that $\mathbf{u} = \begin{pmatrix} -7 \\ 5 \end{pmatrix}$ and $\mathbf{v} = \begin{pmatrix} 2 \\ -3 \end{pmatrix}$, then $\mathbf{u} + \mathbf{v}$ is:
    [A] $\begin{pmatrix} -9 \\ -8 \end{pmatrix}$ [B] $\begin{pmatrix} 9 \\ 2 \end{pmatrix}$ [C] $\begin{pmatrix} -5 \\ 2 \end{pmatrix}$ [D] $\begin{pmatrix} -5 \\ 8 \end{pmatrix}$

30. Two points P and Q have position vectors $\mathbf{OP} = 2\mathbf{i} + 3\mathbf{j}$ and $\mathbf{OQ} = 4\mathbf{i} + 5\mathbf{j}$ then $\mathbf{PQ}$ is:
    [A] 10   [B] $2\sqrt{2}$ [C] 20   [D] $2\sqrt{10}$

31. Given that the vector $x\mathbf{i} + 8\mathbf{j}$ or is parallel to the vector $2\mathbf{i} - 4\mathbf{j}$, then the value of x is:

[A]   −6   [B]   4   [C]   2   [D]   −4
32. A heptagon has _____ sides
   [A]   5   [B]   7   [C]   6   [D]   8
33. The angle x in the figure below is
   [A]   115°   [B]   90°   [C]   65°   [D]   180°

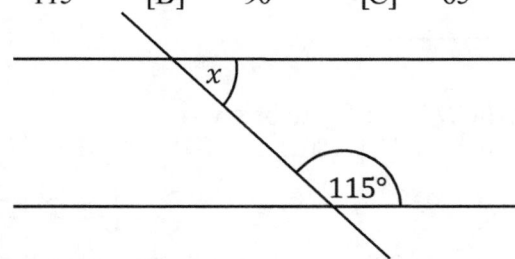

34. The time 5 p.m. on a 24 hour clock is:
   [A]   05:00   [B]   17:00   [C]   10:00   [D]   18:00
35. The value of the angle θ in the figure below is:
   [A]   10°   [B]   60°   [C]   110°   [D]   50°

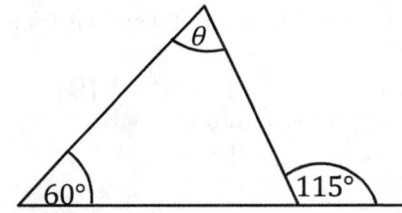

36. The value of $1 - 0.05$ as a fraction is:
   [A] $\frac{19}{20}$ [B] $\frac{1}{20}$ [C] $\frac{9}{10}$ [D] $\frac{5}{10}$
37. The fraction of the area of the rectangle represented by the area of the
   triangle in the figure below is:
   [A] $\frac{1}{8}$ [B] $\frac{1}{4}$ [C] $\frac{1}{16}$ [D] $\frac{1}{32}$

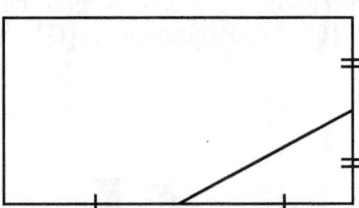

38. The side marked x in the figure below is:
   [A]   18 cm   [B]   13.9 cm   [C]   12 cm   [D]   8 cm

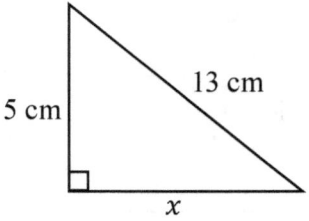

39. The area of trapezium *ABCD* in the figure below is:
    [A] 56 cm   [B] 56 cm²   [C] 80 cm   [D] 80 cm²

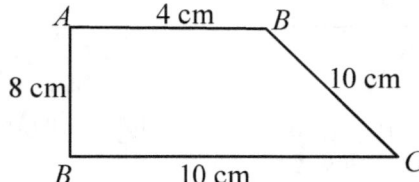

40. The area of a triangle ABC is 12 m². The area of a similar triangle $A'B'C'$ which is 3 times larger is:
    [A]   242 m²   [B] 36 m²   [C]   120 m²   [D]   108 m²

41. The number of lines of symmetry in the figure below is:
    [A] 0   [B] 1   [C] 2   [D] 4

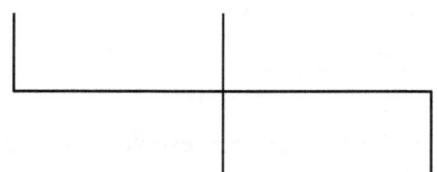

42. The sum of the interior angles of a polygon is:
    [A] 720°   [B] 540°   [C] 630°   [D] 360°

43. The transformation *T*, that has taken place in the figure below is:
    [A]   Rotation [B] Shift [C] Enlargement   [D] Reflection

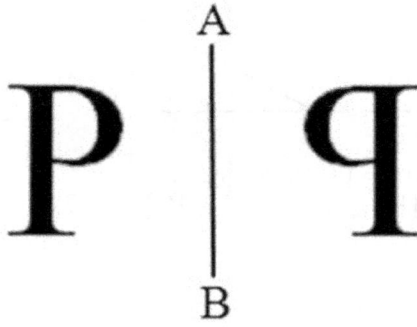

44. The determinant of the matrix $\begin{pmatrix} 6 & 4 \\ 3 & 2 \end{pmatrix}$ is:

    [A]   18   [B]   0   [C]   24   [D]   −24

45. In the figure below, MN is the diameter of the circle with centre O. Angle OPN is:

    [A]   20°   [B]   60°   [C]   70°   [D]   50°

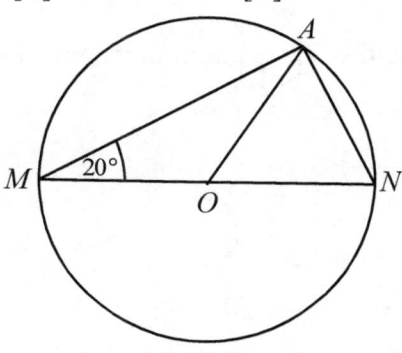

46. The median of the scores 6,3,5,2,6,4,7,4,1,2 is:

    [A]   2   [B]   5   [C]   4   [D]   6

47. The average age of six children is:

    [A]   16 years   [B]   20 years   [C]   14 years   [D]   18 years

48. Below shows a drawn pie chart representing the yield of some fruits. The angle x representing the yield of oranges is:

    [A]   175°   [B]   85°   [C]   90°   [D]   115°

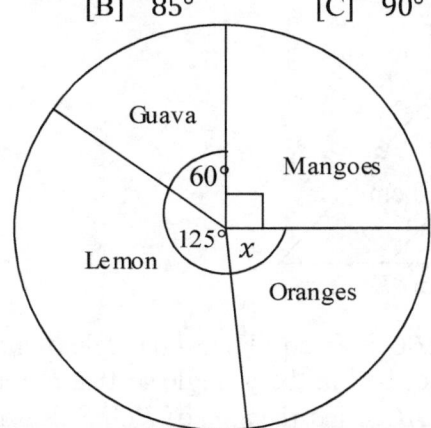

49. Two unbiased diased numbered 1 to 6 are thrown together and their scores recorded. The probability that the sum will be 12 is:

    [A] $\frac{1}{6}$ [B] $\frac{1}{36}$ [C] $\frac{1}{18}$ [D] $\frac{1}{12}$

50. In a box of 50 screws, 4 are faulty. If one screw is taken from the box at random, the probability that it is faulty is:

    [A] $\frac{2}{27}$ [B] $\frac{1}{25}$ [C] $\frac{2}{25}$ [D] $\frac{23}{25}$

STRUCTURAL QUESTIONS 1 - GCE 1984 PAPER 1

Answer all questions

Time allowed: One and a half $\left(1\frac{1}{2}\right)$ hours

1. Given that $\mathbf{u} = \begin{pmatrix} 2 \\ 3 \end{pmatrix}$ and $\mathbf{v} = \begin{pmatrix} 0 \\ 1 \end{pmatrix}$, find numbers a and b such that

    $a\mathbf{u} + b\mathbf{v} = \begin{pmatrix} 4 \\ 5 \end{pmatrix}$.

2. Given that,
    $$A = \{x : x = n+1, n \in N\},$$
    $$B = \{x : x = 2n+1, n \in N\},$$
    $$C = \{x : x = 3n+1, n \in N\},$$

    Draw a Venn diagram to show the relationship between A, B and C. Insert two elements in each region of your diagram.

3. In the figure below, CAD is a right-angle with BE drawn parallel to CD. Given that $BE = 10$ cm, $AE = 6$ cm and $ED = 12$ cm, find
       (a) AB (b) CD.

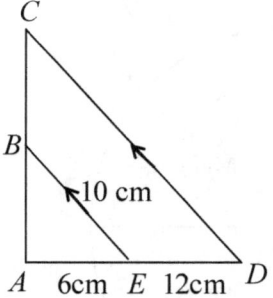

4. In the figure below, ABC is an equilateral triangle of side 20 cm. The rectangle $PQRS$ is inscribed in the triangle so that P and Q are the mid-points of AB and AC. Find the area of $PQRS$, leaving your answer as a surd.

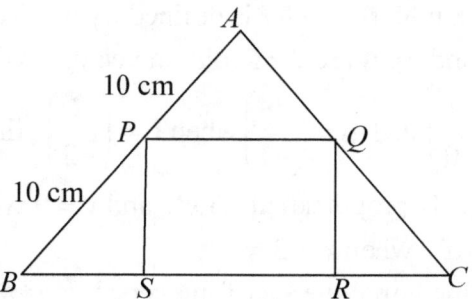

5. In the quadrilateral *OABC*, *D* is the mid-point of *BC* and *G* is the point on *AD* such that *AG* : *GD* = 2 : 1. Given that **OA** = **a**, **OB** = **b** and **OC** = **c** express **OD** and **OG** in terms of **a**, **b** and **c**.

6. In the diagram below *AB* is parallel to *DC* and *DB* bisects angle *ABC*. Given that angle *BAD* = 42°, calculate angle *ABC* and angle *ADB*.

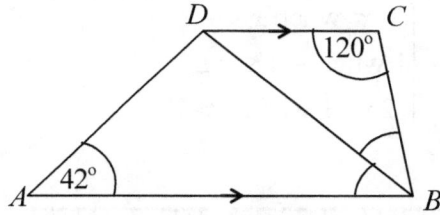

7. The following figure represents a miniature fan which is in the shape of a third of a circle of radius 14 cm. The shaded part is painted and the remainder is a third of a circle of radius 10.5 cm. Taking π as $\dfrac{22}{7}$ calculate the area of the shaded part.

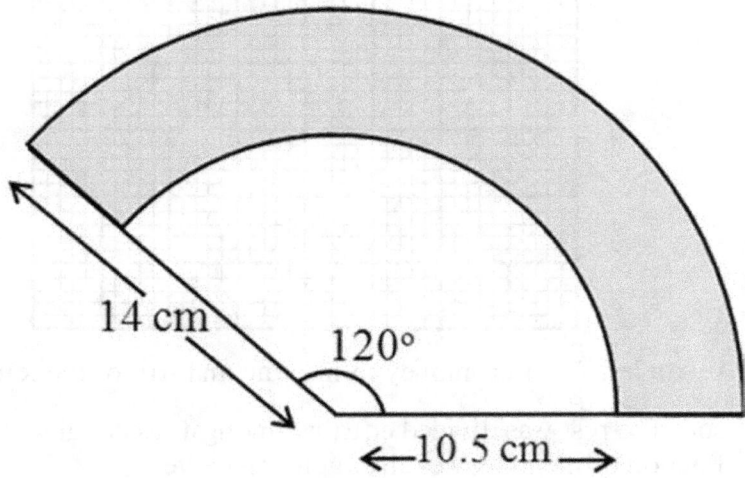

8. A linear transformation $\mathbf{M}: \mathbb{R}^2 \rightarrow \mathbb{R}^2$ is defined by $\mathbf{p} = \mathbf{Mq}$, where $\mathbf{M}$ is 2×2 matrix and $\mathbf{p}, \mathbf{q}$ are 2×1 column vectors. Given that $\mathbf{p} = \begin{pmatrix} 3 \\ 7 \end{pmatrix}$ when $\mathbf{q} = \begin{pmatrix} 1 \\ 0 \end{pmatrix}$ and $\mathbf{p} = \begin{pmatrix} 6 \\ -1 \end{pmatrix}$ when $\mathbf{q} = \begin{pmatrix} 2 \\ -3 \end{pmatrix}$ find $\mathbf{M}$.

9. Given that y is inversely proportional to x^3, and $y = 9$ when $x = 2$, calculate the value of y when $x = 3$.

10. An operation $*$ is defined on $\mathbb{Z}$ the set of integers, by $a*b = a^2b - b^2a$. Evaluate $3*5$ and determine whether the operation $*$ is commutative.

11. The mean of five numbers is 4. When a sixth number is added, the mean of the six numbers is $3\frac{1}{2}$. Find the sixth number.

12. The function $f: R \rightarrow R$ is defined by
$$f: x \mapsto \begin{cases} -x & \text{when } x < -1, \\ 1 & \text{when } -1 \leq x \leq 1, \\ 2x-1 & \text{when } x > 1 \end{cases}$$

Sketch the graph of f.

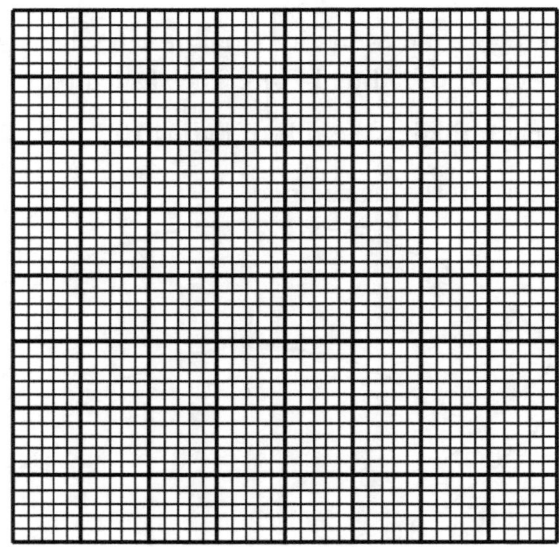

13. A man left $\frac{3}{8}$ of his money to his wife and half of the remainder to his son. The rest was divided equally amongst his daughters. Find what fraction of the money each daughter received.

14. Given that x is an acute angle and $\sin x = \dfrac{5}{13}$, find the value of

    (a) $\cos x$,   (b) $\tan x$

15. Given that $a = 6 \times 10^3$, $b = 2 \times 10^{-4}$, $c = 4 \times 10^{-5}$, find the value of $\dfrac{a \times b}{c}$, expressing your answer in the same standard form.

16. Express w in terms of a, b, u and T, given that $T = \dfrac{wa}{(u+w)b}$.

17. Give, with reasons, what you think are the next two terms in each of the following sequences:

    (a) $1, 3, 7, 13, 21, \ldots$ (b) $1, \dfrac{1}{2}, \dfrac{1}{6}, \dfrac{1}{24}, \dfrac{1}{120}, \ldots$

18. Find the values of m for which the matrix $\begin{pmatrix} 6-m & 2 \\ 25 & 1-m \end{pmatrix}$ is singular, that is, it has no inverse.

19. Divide 303_{five} by 111_{three} and express your result in base five.

20. Given that $r^2 + 3s^2 = 4rs$, find the two values of $\dfrac{r}{s}$.

21. Given that $f : x \mapsto 2x+1$ and $g : x \mapsto 3x^2$. Express fg in the form $fg : x \mapsto \ldots$

22. A student has 3 khaki shirts and 2 white shirts; he also has 3 pairs of khaki trousers and 4 pairs of white trousers. One night in the dark, he chooses one shirt and one pair of trousers at random. Find the probability that the shirt and the trousers are of the same colour.

23. The equation $3x - 5y = 15$ represents a straight line.
    (a) Give one point on this line.
    (b) State the coordinates of the points where the line cuts the x-axis and the y-axis.
    (c) State the gradient of this line.

24. The figure below shows a pyramid on a square base. If all the edges of the pyramid are equal, find, in degrees the value of angle OAC.

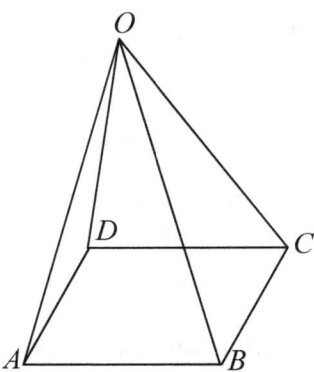

25. An empty bottle weighs 310 g; when full of water 1030 g and when full of alcohol it weighs 880 g. Find the ratio of the weights of equal volumes of alcohol and water.

STRUCTURAL QUESTIONS 2 - GCE 1985 PAPER 1

Answer all questions

Time allowed: One and a half $\left(1\frac{1}{2}\right)$ hours

1. Sets A and B are such that $A\{1.2.,3,4,\}, A \cup B = \{1,2,3,4,5,6\}$.

   List the elements of the set B,
   (a) When B contains the least possible number of elements,
   (b) When B contains the greatest possible number of elements.

2. The area of a circle is 616 cm². Taking π as $\dfrac{22}{7}$ find the radius and the circumference of the circle.

3. Given that $f(x) \equiv (3x+4)(x-1)(x+1)$, find the solution set of the equation f(x) =0,
   (a) in the set of natural numbers,
   (b) in the set of integers,
   (c) in the set of rational numbers.

4. Rewrite the following numbers in ascending order of magnitude,

   $1, -\dfrac{1}{2}, \dfrac{3}{4}, \dfrac{1}{3}, -\dfrac{4}{3}$

5. The nth term u_n of a certain sequence is given by $u_n = (4-n)^2 - 1$. Write down the first five terms of the sequence.

6. The live stock on a certain farm consists of 28 cows, 300 sheep, 74 pigs, 306 poultry, 9 dogs and 3 cats. If this information is recorded on a pie chart, calculate the angle, in degrees at the center of the sector representing the cows.

7. Find the matrix product $\begin{pmatrix} 2 & -3 \\ -2 & 4 \end{pmatrix} \begin{pmatrix} -2 & 1 \\ -5 & 3 \end{pmatrix}$ and calculate its determinant.

8. In the figure below, $OPQR$ is a parallelogram, $TR = 2OT$ and $RM = MQ$. If $\mathbf{OP} = \mathbf{p}$ and $\mathbf{OR} = \mathbf{r}$, express the vectors $\mathbf{TM}$ and $\mathbf{PM}$ in terms of $\mathbf{p}$ and $\mathbf{r}$.

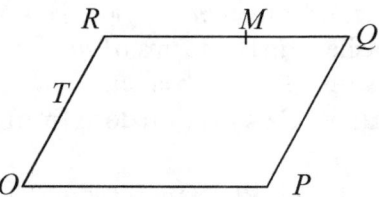

9. The following figure shows the curved surface of a cone formed from the shaded sector of the circle, which subtends an angle of $120°$ at the center. Find the radius, in centimetres, of the base of the cone.

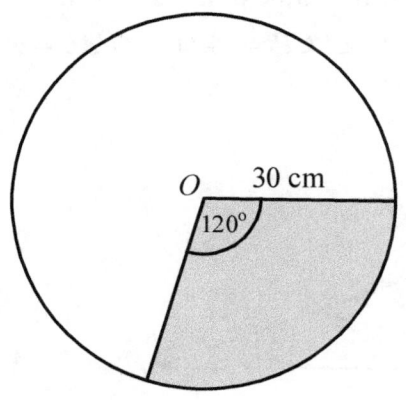
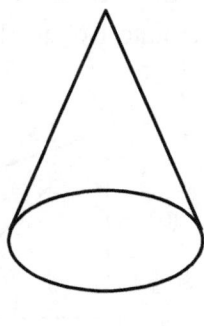

10. Determine which of the following sets consist entirely of elements which are similar figures:
    (a) {triangles}   (b) {squares}   (c) {rectangles}

(d) {parallelograms}   (e) {rhombuses}   (f)   {trapezia}
(g) {hexagons}   (h) {circles}

11. On a map in which 1cm represents 2 km, a plot of land is represented by square of length 2.5 cm.
(a) Calculate the actual area, in km², of the plot of land
(b) If the above scale is written in the form $1:x$, find x.

12. A particle P move so that its speed, v metres per second at time t seconds $(t \geq 0)$, is given by $v = 16t - \frac{1}{4}t^2$

(a) Find the speed of P when $t = 4$.
(b) Show that P starts from rest and find the value of t when p again comes to rest.

13. Decompose each of the following statements into its different components.
(a) Nigeria and Ghana are African countries.
(b) Either Mr. Suh signs the document or he will be dismissed.

14. A basket contains 24 mangoes, of which m are ripe. If 6 more ripe mangoes are added to the basket, find in terms of m,
(a) the number of unripe mangoes in the basket,
(b) the probability that a mango chosen at random from the basket is riped.

15. The triangle with vertices $A(-1,-3), B(2,1)$ and $C(-2,2)$ is

transformed by matrix $\begin{pmatrix} a & b \\ c & d \end{pmatrix}$ to the triangle with vertices

$A(-2,-3), B(4,1), C(-4,2)$. Find the value of a, b, c and d.

16. In figure $AP = 30$ m, $\angle ABP = 48°$, $\angle BCP = 24$, and $\angle PAB = 90°$. Calculate BC, to the nearest 0.1 m.

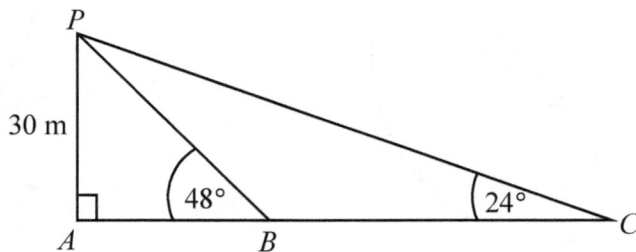

17. In the figure below, ABC is an equilateral triangle, E is the mid-point of AC and BE is parallel to CD. Find, in degrees, the value of x, y and z.

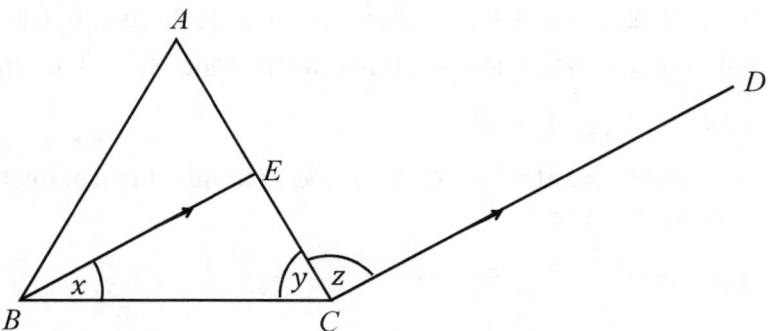

18. Given that $V = \dfrac{\pi n}{v(A-3)}$ express A in terms of V, π and n

19. A certain transformation T is defined by $T:(x,y) \mapsto (2x+y, -x+y)$.

    Find (a) the image of the point $(-2, 6)$ under T,

    (b) the 2×2 matrix representing T.

20. Copy the number line below and on it, illustrate the solution set of the inequality $2 > 1 - x \geq -3$.

21. Shade the region in the figure below which satisfies all three of the following inequalities: $y \geq 0$, $x \geq 0$, $x + y \leq 4$, $2x - y \geq 2$.

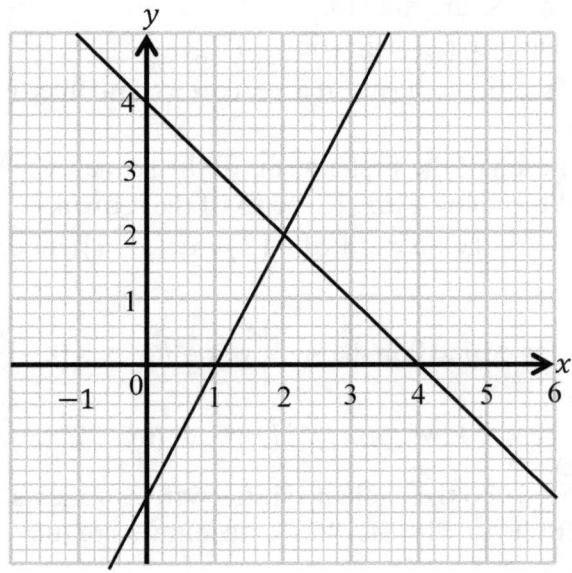

22. Find the remainder when $6x^3 + 27x^2 - 14x + 15$ is divided by $(x+5)$.

23. A formula for converting temperatures from Celsius $C°$ to Fahrenheit $F°$ is given by $F = \dfrac{9}{5}C + 32$.

   Find the temperature at which Celsius and Fahrenheit thermometers record the same value.

24. Given that $\mathbf{M} = \begin{pmatrix} 4 & 2 \\ 6 & 1 \end{pmatrix}$, find $\mathbf{M}^{-1}$.

25. Evaluate, without using tables or calculators, $\left(\dfrac{0.027}{64000} \right)^{\frac{1}{3}}$, giving your answer in decimals.

STRUCTURAL QUESTIONS 3 - GCE 1987 PAPER 1

1. Given that the mean of 3, 4 and m is 6, find the mean of 2, m and 14.
2. A dealer sells a motorcycle for 780,000 FCFA and makes a profit of 30 % on his cost price. Find the cost price.
3. The velocity of light is approximately 300,000 km/s. Express this velocity in m/s, in standard.
4. Find the value of x for which $62_x = 44_{ten}$
5. The sum of the first n terms of a sequence of number is given by $3n^2 + 2n.$
   (a) Find each of the first 3 terms of the sequence.
   (b) Find the sum of first 12 terms of the sequence.
6. Given that $5x^2 + 4x + 3 \equiv a + b(x+1) + cx(x+1)$, find the constants a, b and c.

7. A function f is defined on the set $\left\{ x : x \in R, \quad x \neq 0, -3 - \dfrac{10}{3} \right\}$ by

   $f : x \mapsto \dfrac{1}{x+3}$. Find (a) $f^{-1}(x)$ (b) $ff(x)$.

8. In September 1986 the exchange rates were as follows:
   $\$1 = 340\mathbf{FCFA}$, £1 $= \$1.146$. A student was required to pay £ 2000

as tuition for the 1986/87 academic year in a British university. Determine how much money was required in FCFA.

9. Find the set of real values of x for which $x^2 + 2x - 3 < 0$.

10. The figure below shows the graph of three straight lines.

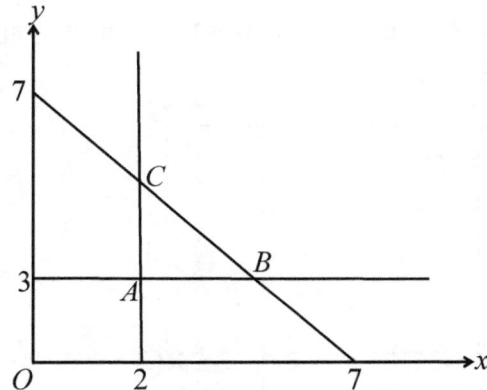

Write down the equations of these straight lines:

AB...

AC...

BC...

Write down the inequalities which define points inside triangle *ABC*.

11. Given that $a = p\left(1 + \sqrt{\dfrac{r}{t}}\right)$, Express t in terms of a, p and r

12. Given that $\log(8) = 0.9030$, write down, without the use of tables, the values of (a) $\log(64)$ (b) $\log(2)$ and (c) $\log(\sqrt{2})$.

13. The figure below shows a circle with centre O and four points A, B, C and d on the circle. ON is perpendicular to AB.

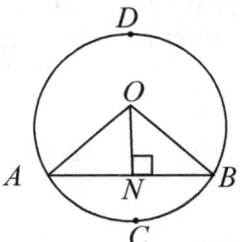

Give a name for each of the following:

(a) the set of points on ACB

201

    (b)   the set of points on OA
    (c)   the triangle OAB
    (d)   the set of points on AB
    (e)   the set of points in the region $OACBO$

14. The figure below shows a cube. In the questions that follow, only one of the alternatives is true.

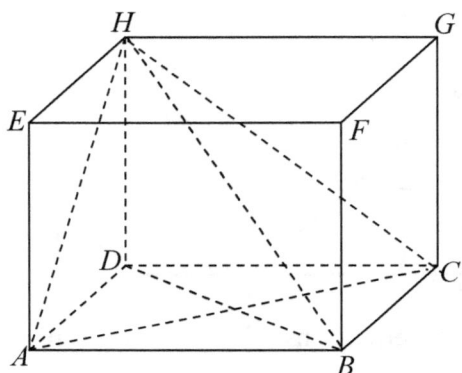

Mark a circle around the correct alternative.

(i) ΔHDB is:

       [A] isosceles but not equilateral
       [B] Right-angled and scalene
       [C] Equilateral
       [D] Scalene but not equilateral

(ii) ΔHAC is:

       [A] isosceles but not equilateral   [B] Right-angled.
       [C] Equilateral.   [D] Scalene.

(iii) ΔHAD is congruent to:

       [A] ΔHDB [B] ΔADC [C] ΔAHC [D] ΔBHC

(iv) $\angle HAC$ is

       [A] $30°$ [B] $45°$ [C] $60°$ [D] $90°$

15. Two quantities x and y are believed to be connected by the relation

$$y \propto \frac{1}{x^2}.$$

In an experiment, x and y were measured and the following results obtained.

x	2	3	4	6	8	12
y	36	16	9	4	3	1

It is suspected that one of the values of y was wrong. Find the wrong one and give its correct value.

16. Given that $s = \dfrac{T}{5} - \dfrac{2R}{W}$. Find s when $T = 10.5$, $R = 5.5$ and $W = 10$.

17. If $A = \{x : -4 \leq x < 6\}$ and $B = \{x : x < 0\}$,

    Illustrate $A \cap B$ on the following number line:

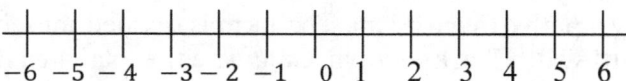

18. In the figure below, $AB = 10$ cm, $PB = 3$cm, $BC = 20$ cm and
    $AN = 9$ cm. AN is perpendicular to BC. If PQ is parallel to BC.
    Calculate
    (a) the length of PQ,   (b) the area of triangle APQ.

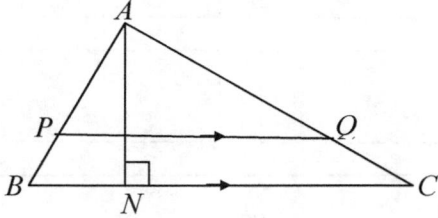

19. Find the values of x and y in the following figure showing your working clearly.

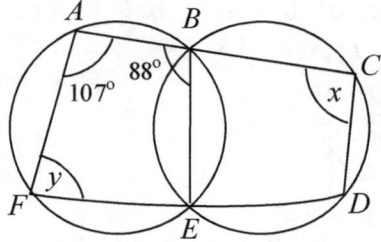

20. Determine which of the following matrices is/are singular
$$\begin{pmatrix} -2 & 2 \\ 2 & 2 \end{pmatrix}, \begin{pmatrix} -2 & -2 \\ 2 & 2 \end{pmatrix}, \begin{pmatrix} -2 & -2 \\ 2 & -2 \end{pmatrix}, \begin{pmatrix} -2 & -2 \\ -2 & 2 \end{pmatrix}, \begin{pmatrix} 2 & -2 \\ -2 & -2 \end{pmatrix}.$$

21. If $\mathbf{a} = \begin{pmatrix} 2 \\ 5 \end{pmatrix}$, $\mathbf{b} = \begin{pmatrix} 4 \\ 9 \end{pmatrix}$, $\mathbf{c} = \begin{pmatrix} 1 \\ 3 \end{pmatrix}$. Find a relationship between $\mathbf{a}$, $\mathbf{b}$, and $\mathbf{c}$,

    in the form $\mathbf{b} = u\mathbf{a} + v\mathbf{c}$, where u and v are integers.

22. Given the straight lines

$$A : 2x = 3y + 5, \quad B : 2y = -4x + 3,$$
$$C : 2x + y = 5, \quad D : 3x + 2y = 3,$$

Determine which of the lines (if any)

(a)   are parallel
(b)   are perpendicular
(c)   pass through the origin
(d)   pass through the point(2,1)

23. The figure below shows a rectangular farm 80 m long and 30 m wide, surrounding by a path of width 1 m. The farm is divided into four plots by paths of width 1 m as shown. Calculate the total area of all the paths.

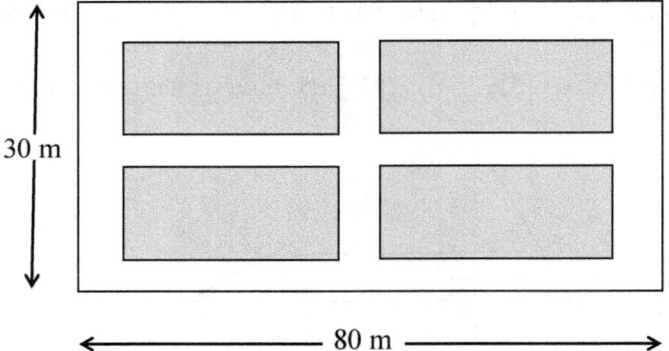

30 m

80 m

24. Given that $I = \{$positive integers, excluding zero$\}$, list the four smallest positive elements in each of the sets A, B, C, where

$$A = \{2x + 1, x \in I\},$$
$$B = \{2x^2 - 1, x \in I\},$$
$$C = \{(-1)^x x, x \in I\}.$$

25. On a graph paper,   draw a
   (a) Parallelogram   (b) rhombus   (c) trapezium   (d) kite

ESSAY QUESTIONS 1 - NWR MOCK 2015 P2

Answer all questions

Time allowed: Two and a half $\left(2\frac{1}{2}\right)$ hours

SECTION A

1. Simplify $2\frac{7}{9} \times \frac{3}{5} - 2\frac{1}{6} \div \frac{1}{4}$.

2. Let p and q be two statements defined as follows.
   p: John is hungry.
   q: John is thirsty.
   Write out the following in ordinary English language avoiding any technical words such as "negation", "implies" etc
   (i) $\sim p$: _____

   (ii) $p \wedge q$: _____

   Write the following in symbolic language.
   (iii) John is neither hungry nor thirsty_____

   (iv) Whenever John is hungry, he is thirsty_____

3. Find the remainder when $f(x) = x^3 + 4x^2 + 5x + 2$ is divided by $(x + 1)$.
   What conclusion do you draw?

4. In the figure below PQR is a secant while TR is a tangent to the circle at O. Given that $TR = 6$ cm, $QR = 5$ cm and $PQ = x$ cm. Calculate the value of x.

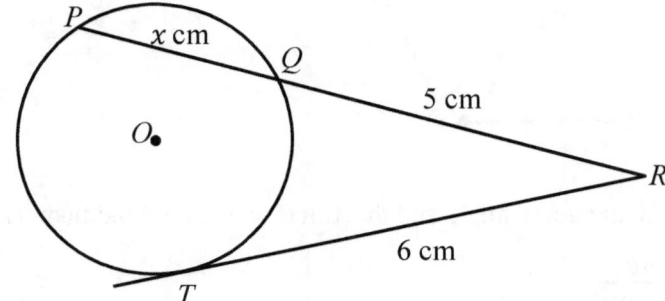

5. The table below shows the frequency of visits of some students to Lake Oku during a term.

Number of visits	0	1	2	3	4	5	6
Number of students	5	8	5	6	3	2	1

(a) State the modal number of visits.

(b) Find the median number of visits.

6. The nth term of an Arithmetic progression is $12 - 4n$. Find the common difference.

7. Find the values of x and y in the figure below.

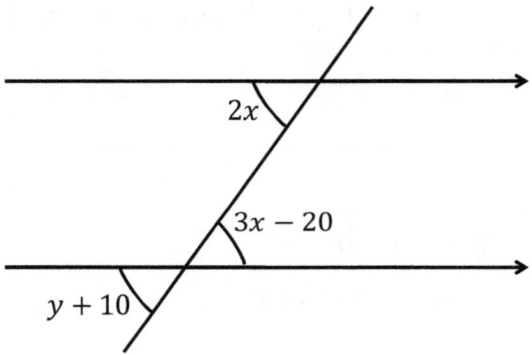

8. Solve and represent on a number line the inequality $(x - 1)(x + 5) > 0$.

9. Given the network below and the diagraph $D = (V, A)$. List the elements of A.

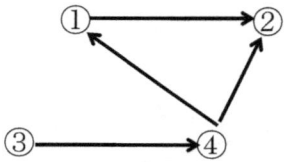

10. Given that θ is and acute angle and that $\tan \theta = \dfrac{7}{24}$. Find the numerical value of $\dfrac{\sin \theta}{\sin \theta + \cos \theta}$.

11. Given the points $O(-3, 2)$ and $P(4, -6)$.

    (a) Find the coordinates of the midpoint $M(x, y)$ of $[OP]$.

    (b) The distance between the points O and P.

12. Solve the system of equations $\begin{cases} 3x - y = 5 \\ x + y = 7 \end{cases}$

13. The following figure illustrates the relationship "is younger than" between 5 students.

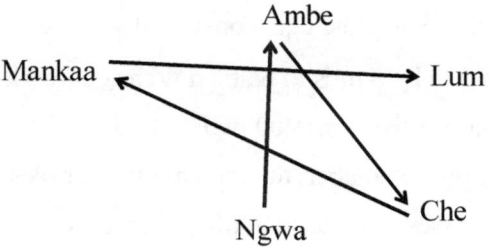

Use the figure to identify who is:

(a) The youngest student. (b) The youngest student.

14. Find the value of x for which $2^x = \frac{1}{8}$.

15. Given the vectors $\mathbf{p} = \begin{pmatrix} 2 \\ -1 \end{pmatrix}$ and $\mathbf{q} = \begin{pmatrix} 2 \\ -6 \end{pmatrix}$. Find $3\mathbf{p} + \mathbf{q}$.

SECTION B

1. (i) Three girls each bought a cooker of different mark. Each of them had a discount of 11% on the respective displayed prices. The total amount of discount for all of them was 84,150 FRS. Given that the discount on each cooker was proportional to the life span of the cooker, which is 54 months, 72 months and 78 months respectively.

(a) Calculate the discount granted on each of the cookers.

(b) Determine the displayed price on the cheapest of the cookers.

(c) Determine how much the girl who bought the cheapest cooker paid.

(ii) Consider the polynomial $P(x) = (4x + 5)(x + 2) - (x + 2)^2$.

(a) Factorize $P(x)$ completely.

(b) Expand, simplify and arrange $P(x)$ in the form,
$P(x) = ax^2 + bx + c$, where $a, b, c \in \mathbb{Z}$.

(c) Solve in $\mathbb{R}$ the equation $P(x) = 0$.

2. The function f and g are defined as $f: x \mapsto x - 1$; $g: x \mapsto x^2 - 4$.

    (a)   Find $g(-7)$ (b)   Find $gf(x)$

    (c)   Construct a table of values for $gf(x)$ for which $-3 \leq x \leq 4$.

    (d)   Taking 1 cm for 1 unit on both axes, draw the graph of $y = gf(x)$.

    (e)   Using your graph solve the equation $y = 0$.

  (ii) A room measuring 5 m by 6 m is to be tiled with square tiles of size 25 cm each. Given that one tile costs 400 FCFA, find

    (a) To the nearest whole number, the total number of tiles to be used.

    (b) The total cost of tiles to be used in tiling this room.

3. (i) The following is a Venn diagram.

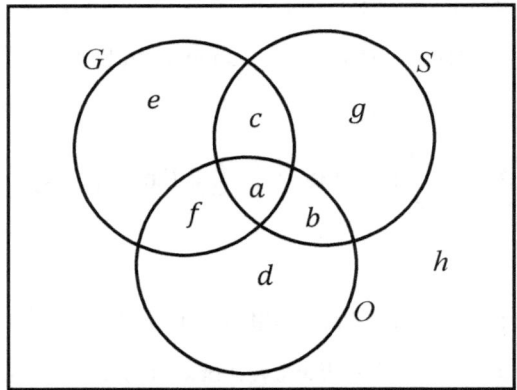

Let $\mathscr{E} = \{x : x$ is a student in a college$\}$

    $G = \{x : x$ is a girl$\}$

    $S = \{x : x$ is a student who can swim$\}$

    $O = \{x : x$ is a student over 15 years of age$\}$.

The eight regions formed by the intersection of these sets are marked a to h.

Which regions best represents the following sets of students?

    (a) $\{x : x$ is a girl over 15 years of age who cannot swim$\}$.

    (b) $\{x : x$ is a boy over 15 years of age who can swim$\}$.

    (c) $\{x : x$ is a girl under 15 years of age who cannot swim$\}$.

    (d) $\{x : x$ is a boy under 15 years of age who can swim$\}$.

    (e) $\{x : x$ is a student under 15 years of age who can swim$\}$.

(ii) The number of questions answered by a student in a mathematics examination is directly proportional to the square root of the time taken in minutes. Given that the student answered 2 questions in 36 minutes.

(a) Find an equation linking the time and the number of questions answered.

(b) Find the constant of proportionality.

(c) Find the time taken to answer 5 questions.

4. (i) The marks scored by 40 students in a particular examination are as follows.

38	74	28	32	10	31	49	34	50	39
38	92	50	42	38	64	24	65	9	77
18	35	12	87	41	27	8	90	22	21
42	43	52	51	72	70	90	91	29	28

(a) Prepare a group frequency table for this data using class intervals; 1-20, 21-40, 41-60, ….

(b) Use the table to draw the histogram for this data.

(c) Calculate the mean mark.

(ii) A bag contains 10 balls that differ only in colour with 4 blue and 6 red. Two balls are picked one after the other with replacement. What is the probability that;

(a) Both balls picked are red.

(b) Both balls picked are of the same colour.

ESSAY QUESTIONS 2 - CGCE JUNE 2004 PAPER 2
Answer all questions
Time allowed: Two and a half $\left(2\frac{1}{2}\right)$ hours

1. (i) The functions f, g and h are defined on $\mathbb{R}$, the set of real numbers as:
$$f: x \mapsto x^2 - 3, g: x \mapsto x + 2, \ h: x \mapsto px + q,$$
Where p and q are constants.
(a) Evaluate $f(-2)$

(b) Find the composite function $f \circ g(x) = \cdots$

(c) Express g^{-1} in the form $g^{-1}: x \longmapsto \cdots$

(d) Given that $g \circ h(x) = 3x + 1$, dtermine the value of p and q.

(ii) Solve for x and y, the simultaneous linear equations:
$$4x - 3y = 4$$
$$2x - y = 6$$

2. (i) Given the statements:

p: Bih is lazy

q: Bih is beautiful.

Write down the following using symbolic language.

(a)   Bih is not lazy.

(b)   Bih is lazy and beautiful.

(c)   Bih is neither lazy nor beautiful.

(d)   Bih is either lazy or beautiful.

(e)   Bih is lazy but not beautiful.

(f)   It is not true that Bih is lazy and beautiful.

(ii) Given that $\mathscr{E} = \{x: x \text{ is a student in form five}\}$.

$H = \{x: x \text{ is a student who does History}\}$

$G = \{x: x \text{ is a student who does Geography}\}$

$B = \{x: x \text{ is a student who does Biology}\}$

Write the following statements in ordinary English, avoiding the use of technical words such as union, intersection, etc:

(a) $B \cap H = \emptyset$ (b) $B \subset G$

Write the following in set notation:

(c) All students study Geography

(d) No student does all the three subjects

(e) Forty students study both History and Geography.

3. (i) In the figure below, T is the midpoint of AB and M is the midpoint of AT. Given that $\mathbf{OA} = \mathbf{a}$ and $\mathbf{OB} = \mathbf{b}$

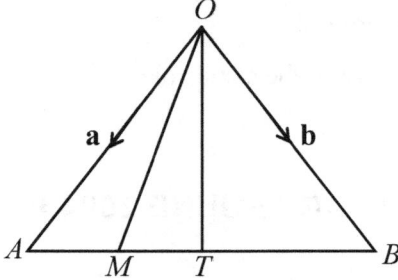

Express as simply as possible, in terms of $\mathbf{a}$ and $\mathbf{b}$ the vectors.

(a) $\mathbf{AB}$ (b) $\mathbf{AM}$ (c) $\mathbf{OM}$

(ii) Given three vectors as: $\mathbf{p} = \begin{pmatrix} 3 \\ 2 \end{pmatrix}$, $\mathbf{q} = \begin{pmatrix} 5 \\ 4 \end{pmatrix}$ and $\mathbf{r} = \begin{pmatrix} -2 \\ -1 \end{pmatrix}$

(a) Write as a column vector $2\mathbf{p}$

(b) calculate |**q**| the magnitude of **q**.

(c) Express $2\mathbf{p} - \mathbf{r}$ in the form $m\mathbf{i} + n\mathbf{j}$, where I and j are unit vectors along the positive x- and y-axes respectively.

(d) Given that $t\mathbf{p} + k\mathbf{r} = \mathbf{q}$. Determine the values of t and k, where t and k are constants.

4. (i) 36,000 people watched a football match between two teams, Kumbo strikers and Victoria United. $\frac{3}{5}$ of the people supported kumbo strikers. 4800 people supported none of the teams and the rest of the people supported Victoria united. Determine

(a) The number of people who supported Kumbo Strikers.

(b) The percentage of the crowed that supported Victoria united. (Give your answers to 1 d.p.)

The total gate takings at the match amounted to 12,600,000 FCFA. This amount was shared amongst FECAFOOT, Match officials, Kumbo strikers and Victoria United in the ratio 8: 2: 5: 5. Find

(c) The average amount paid by each spectator at the gate.

(d) The amount paid to Kumbo strikers.

(ii) An athlete completes a race of 500 m in 55 seconds. The athlete starts off at an average speed of 8m/sec and runs for t seconds. Then runs the rest of the distance at an average speed of 10 m/sec. Calculate the distance covered in the first part of the race.

5. (i) A boy measured to the nearest metre how far he could through a tennis ball on 20 successive trials and obtain the following results:

66	69	70	68	71	68	69	70	67	68
68	68	67	66	69	68	69	70	68	67

(a) Draw a frequency distribution table for the data.

(b) State the mode of the distribution.

(c) Find the median of the distribution.

(d) An approximate formula for determining the mean is given by
mean − mode = 3(mean − median)
Use this formula to calculate the mean of the distribution.

(e) Calculate the exact mean of the distribution to one decimal place.

(ii) Andrea has 6 white panty hoses and 10 black pantyhoses of the same kind in a drawer. In the dark, she takes two pantyhoses, at random one after the other without replacement. Calculate, the probability that she takes out

(a) A white pantyhose in the first draw.

(b) A black pantyhose in the second draw.

(c) A white and a black pantyhose in that order.

6. (a) Complete the table below, for the function $f(x) = 2x + \frac{1}{x}$.

x	0.25	0.5	1.0	1.5	1.0	1.5	2	2.5
$f(x)$	4.5							5.4

(b) Using 2 cm to represent 1 unit on the y-axis and 4 cm to represent 1 unit on the x-axis, draw the graph of $f(x) = 2x + \frac{1}{x}$.
Use your graph to answer the following questions.

(c) Write down the minimum value of $f(x)$ within the range of values 0.25 to 2.5.

(d) Find the gradient of the curve at the point where $x = 0.5$.

(e) Solve for x, the equation $f(x) = 3.5$, give your answer to 2 decimal places.

7. The triangle with vertices $A(1,0)$, $B(2,1)$ and $C(4,0)$ is mapped uonto triangle A', B', C' by the transformation $T = \begin{pmatrix} 0 & 1 \\ -1 & 0 \end{pmatrix}$.

   (a) Find the coordinates of the images A', B' and C'.

   (b) Using the scale of 2 cm to 1 unit on both axes, plot the triangle ABC and its image $A'B'C'$ on graph paper.

   (c) Describe the transformation T geometrically.

   (d) Determine the inverse of the transformation matrix. Hence or otherwise, describe completely the transformation represented by this inverse matrix.

8. (i) A ship sails 15 km due east from a harbour H to a point P. It then sails due south to a point Q, where it stops and anchors. The distance from P to Q is 25 km.

   (a) Draw a diagram to show the ship's course from H to Q.

   (b) Calculate to one decimal place how far the ship is from its starting point.

   (c) Find to the nearest whole degree, the bearing of H from Q.

   (ii) Solve for x, the following equations where $0° \le x \le 180°$.

   (a) $\cos x° = 1$ (b) $\sin x° = \cos 15°$.

   (iii) Given that PQ is parallel to BA. Find the values of the angles x, y and z in the following figure.

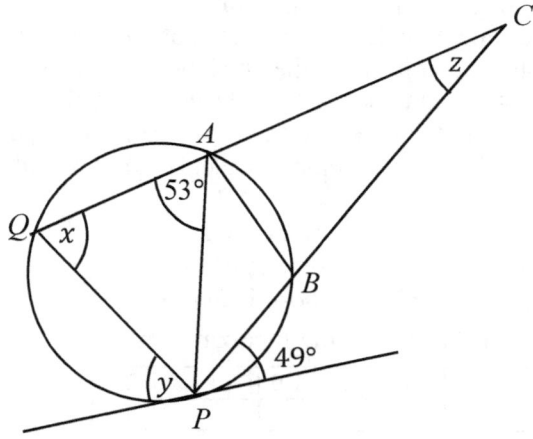

9. (i) In the figure above AB is parallel to DC, $AB = 32$ cm, $BE = 12$ cm and $DE = 3$ cm.
   (a) Show that the triangles ABE and CDE are similar.
       Calculate
   (b) The length CD.
   (c) The ratio of the areas of triangle ABE to trangle CDE .

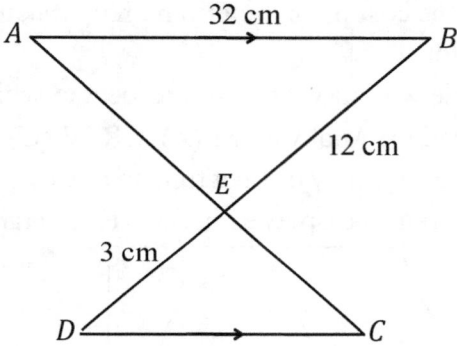

(ii) Given that $f(x) = x^3 + 6x^2 + 5x - 12$.
   (a)   Evaluate $f(1)$.
   (b)   Hence factorize $f(x)$ complete.

ESSAY QUESTIONS 3 - CGCE JUNE 2005 PAPER 2

Answer all questions
Time allowed: Two and a half $\left(2\frac{1}{2}\right)$ hours

1. (i) A palm wine seller buys overnight palm wine at 60 FCFA a litre and fresh palm wine at 80 FCFA a litre. She mixes the overnight and fresh wines in the ratio 2:3 before selling to customers. Given that she sells 100 litres of the mixture daily at 125 FCFA per litre,
   (a) Find the number of litres of fresh palm wine in the mixture that she sells daily.
   (b) Find the total cost of 100 litres of the mixture.
   (c) Find also her total profit for the day.

(ii) A complete wedding dress in a shop had a price tag of 50.000 FCFA on it. The shopkeeper sold it to a bride-to-be, in cash at a discount of 16%.

    (a) Calculate how much the bride-to-be paid for it.

    (b) In selling the dress at the discounted price the shopkeeper made a profit of 20% on the cost price; determine how much the shopkeeper paid for it.

2. A sample of 116 people was surveyed with respect to which of the 3 radio stations they tuned to: Abakwa FM (A), CRTV (C) and Afrique Nouvelle (N) 40 people said they did not tune to any of the 3 stations. The rest of the information is displayed in the Venn diagram below.

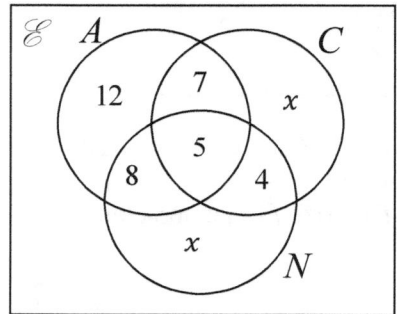

    (a) State how many tuned only to Abakwa FM.

    (b) State how many tuned to both Abakwa FM and CRTV.

    (c) Determine the value of x.

    (d) Find how many tuned to more than one station

    (e) Determine how many tuned to exactly 2 stations.

    (f) Copy the Venn diagram and shade on it the region corresponding to $(A \cup N) \cap C'$.

    (g) Describe in ordinary English and without using technical terms the set given by $(A \cup N) \cap C'$.

3. An incomplete table of values is given below for the function $f(x) = 2 + 3x - x^2$.

x	-2	-1	0	1	2	3	4	5
y	-8			4	4			

    (a) Copy and complete the table.

Taking 1cm to represent 1 unit on the *y*-axis and 2 cm to represent 1 unit on the *x*-axis, draw the graph of $f(x)$.

(b) Use your graph to find the maximum value of $f(x)$ and the corresponding value of *x*, correct to 1 decimal place

(c) By drawing a suitable straight line on the same axes, solve for *x*, the equation $2 + 3x - x^2 = 0$.

4. (i) Find the sizes of the angles marked *x*, *y* and *z* in figure (a) and (b) below.

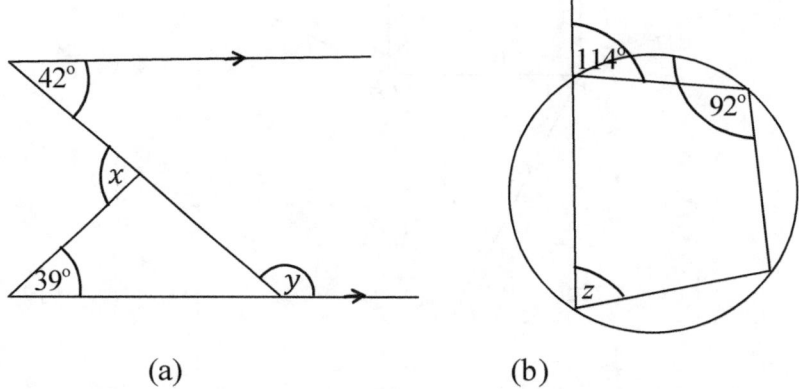

(a)   (b)

(ii) The function, $f(x)$ where *p* and *q* are constants; has $x + 2$ as a factor and leaves a remainder 60 when divided by $x - 2$.

    (a) Find the values of *p* and *q*.

    (b) Find the other factors of $f(x)$.

5. (i) The following figure is the net (not drawn to scale) of a pyramid on square base *ABCD* of side 6 cm. The slant edges *FA*, *FB*, *FC*, and *FD* are equal and of length 5cm.

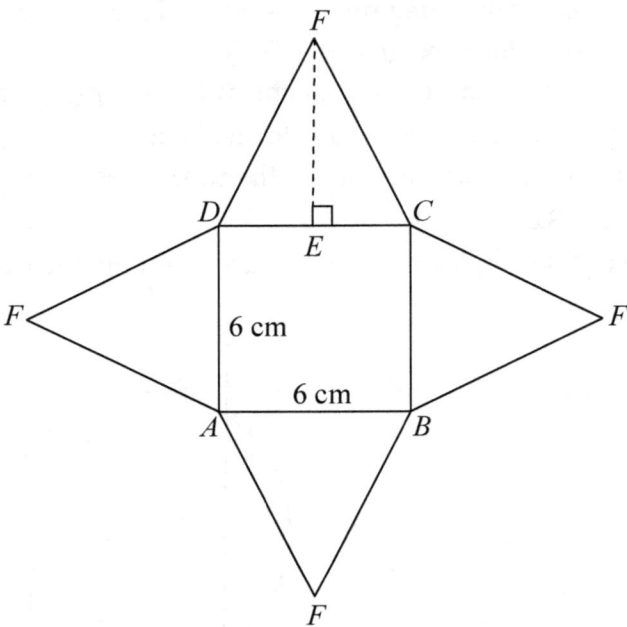

State:
  (a) The number of faces,
  (b) The number of edges, and
  (c) The number of vertices possessed by the pyramid
  (d) Calculate the altitude, FE, of FDC and the total surface area of the pyramid.

(ii) Give that the matrix $\mathbf{M} = \begin{pmatrix} 4 & 3 \\ 3 & 2 \end{pmatrix}$, find the matrix $\mathbf{N}$ such that

$\mathbf{MN} = \mathbf{NM} = \mathbf{I} = \begin{pmatrix} 1 & 0 \\ 0 & 1 \end{pmatrix}.$

  Hence, or otherwise find the values of a and b such

  that $\begin{pmatrix} 4 & 3 \\ 3 & 2 \end{pmatrix} \begin{pmatrix} a & b \\ -9 & -10 \end{pmatrix} = \begin{pmatrix} 1 & 2 \\ 3 & 4 \end{pmatrix}.$

6. The functions f, g and h are defined on $\mathbb{R}$, the set of real numbers, thus: $: x \mapsto 2x - 1, x \in \mathbb{R}$, $g : x \mapsto x^2$, $x \in \mathbb{R}$
   $h : x \mapsto \dfrac{1}{4-x}$, $x \in \mathbb{R} - \{4\}$.

(a) Evaluate $f(3)$ and $h(-2)$

(b) Express in the form above, the functions h^{-1} and fg stating the domain of definition for each function.

(c) Solve, for x the equation $g \circ h(x) = 1$.

7. Use a pencil, ruler and a pair compasses only in this question.

   (a) Draw a circle with centre O and radius 5 cm

   (b) Construct a chord AB of length 8 cm

   (c) At the point A construct angle $DAB = 60°$, where D is a point on the circumference of the circle.

   (d) Determine, by construction, the point C on the minor are DB such that $DC = CB$

   (e) Draw the lines DC and CB

   (f) Measure the length of AD correct to 1 decimal place

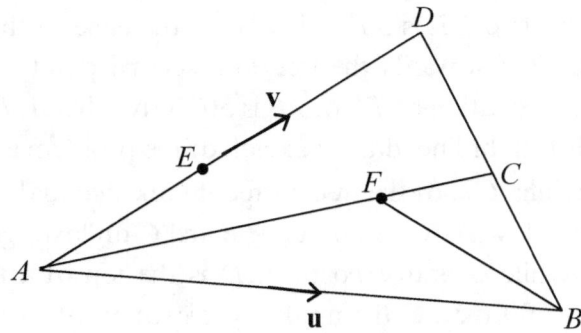

In the figure above, C and E are the midpoints of DB and AD respectively. The point F divides AC in the ratio 2: 1. Given that $\mathbf{AB} = \mathbf{u}$ and $\mathbf{AD} = \mathbf{v}$.

(a) Express, in terms of $\mathbf{u}$ and $\mathbf{v}$, the vectors $\mathbf{BD}$ and $\mathbf{AF}$

(b) Show that $\mathbf{BF} = k\mathbf{BE}$, where k is a constant.

7.8	8.1	8.3	8.0	7.8	8.0	7.9	8.0	8.2	8.1
8.1	7.7	8.0	7.9	8.0	7.9	8.2	8.1	7.8	8.0
8.2	8.1	7.9	8.0	8.1	7.9	8.0	8.3	8.1	8.0
8.4	8.1	8.4	8.1	8.4	8.0	8.2	8.0	8.2	8.1
8.0	7.8	8.1	8.2	8.3	7.9	8.1	7.7	8.0	8.3

8. (i) The table above shows the lengths of 50 counting sticks measured in cm, to 1 decimal place. Copy and complete the table below.

Length, x (cm)	7.2	7.8	7.9	8.0	8.1	8.2	8.3	8.4
Tally								
Frequency								

(a) State the modal length.

(b) Determine the median length to 2 decimal places.

(c) Calculate the mean length, expressing your answer correct to 2 decimal places.

(ii) A bag contains 1 white ball and 2 red balls, all the balls being identical but for colour. A trial consists of drawing a ball, noting the colour and putting it back in the bag and drawing a second ball. By drawing a tree diagram or otherwise, find the probability that the second ball drawn is red.

9. (i) From a point A, a surveyor notes that he angle of elevation of the top, T, of a palm tree CT is 30°, where C is the base of the tree. The surveyor walks 50m towards the tree to a second point B. He notes that the angle of elevation of T from B is 50°. Give that A, B and C are in a horizontal straight line, draw a sketch of the problem and use it to calculate the height, CT, of the tree correct to one decimal point.

(ii) The figure below shows 3 points A, B and C on level ground. B is due east of A while C is due north A. D is the top of a tower, CD. Give that $AC = 500$ m, and the angles of elevation of D from and B are 30° and 25° respectively.

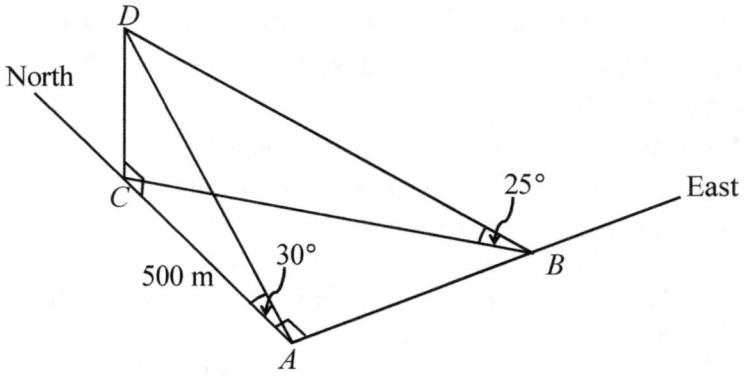

(a) Calculate the height, CD, of the tower, to 2 decimal places

(b) Hence, calculate the distance BC to 2 decimal places.

(c) Calculate the bearing of C from B.

ESSAY QUESTIONS 4 - CGCE JUNE 2006 PAPER 2

Answer all questions

Time allowed: Two and a half $\left(2\frac{1}{2}\right)$ **hours**

1. (i) A television set costs 220,000 FCFA on cash payment. Lucy bought the television set on hire purchase thus:
    - she first paid a deposit of 20% of the cash payment price followed by
    - 33 monthly instalments of 5,500 FCFA
    a) Find the amount she paid as deposit
    b) Find the total amount she paid for the television set
    c) Determine how much more she actually paid for the television set
    d) Hence, express this extra cost as a percentage of the cost on cash payment. (Express your result to 1 decimal place)

    (ii) A number x is such that the number squared plus three times the number is equal to 18.
    a) Form an equation in terms of x.
    b) Hence, solve, for x, the equation in (a) above.

2. (i) Given that $\begin{pmatrix} 3 & 1 \\ 1 & 0 \end{pmatrix}\begin{pmatrix} 4 & 9 \\ x & 2 \end{pmatrix} = \begin{pmatrix} 15 & y \\ z & 9 \end{pmatrix}$ find the values of x, y and z.

    (ii) Given that matrices $\mathbf{A} = \begin{pmatrix} 3 & -2 \\ 4 & -4 \end{pmatrix}$ and $\mathbf{B} = \begin{pmatrix} 5 & 6 \\ -2 & 3 \end{pmatrix}$

    Find
    (a) The transpose of $\mathbf{B}$ (b) The inverse of $\mathbf{A}$ (c) $3\mathbf{A} - \mathbf{B}$

    (iii) The network connecting towns, P, Q and R shown in the figure below can be expressed as a route matrix, $\mathbf{M}$, where the entries show the number of routes leading to the town.

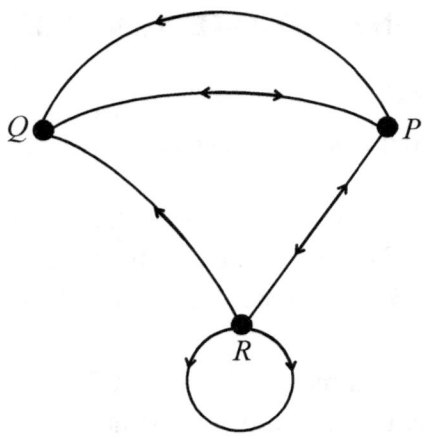

$$\begin{array}{c} \text{To} \\ \begin{array}{ccc} P & Q & R \end{array} \\ \mathbf{M} = \text{from} \begin{array}{c} P \\ Q \\ R \end{array} \left(\begin{array}{ccc} & 2 & \\ & 0 & \\ 1 & & 2 \end{array} \right) \end{array}$$

Complete the entries in the matrix, **M**

3. (i) In a birthday party 3 main types of drinks were served, namely: Wine (*W*) Beer (*B*) and Top (*T*) Each of the 80 guests at the party drank at least one type of drink. The information below shows the distribution of what guests drank

   x drank top only

   $x + 2$ drank beer only

   $x + 3$ drank wine only

   10 drank wine and top only

   9 drank wine and beer but not top

   17 drank wine and beer

   23 drank beer and top

   a) Draw a Venn diagram to represent the above information
   b) Find the value of x
   c) Find the number of guests who drank wine only

d) Calculate the number of people who drank beer and top but not wine

(ii) Describe, in set notation, the relationship that exists between the shaded regions and the sets in each of the diagrams below.

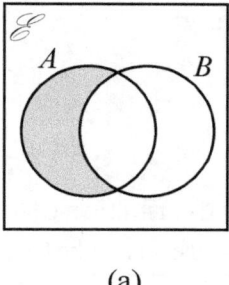   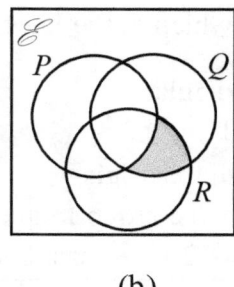   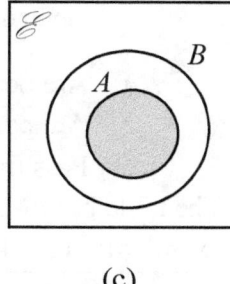

(a)   (b)   (c)

4. (i) Given the functions $f(x) = x^2$, $x \in \mathbb{R}$, $g(x) = \frac{1}{x-4}$, $x \in \mathbb{R}$, $x \neq 4$ and $h(x) = kx + m, x \in \mathbb{R}$.

(a) Expain the meaning of the statement "$x \neq 4$" in the function $g(x)$. Find

(b) $g(0)$ (c) $g^{-1}(1)$

(d) The values of k and m given that $f \circ h \ (x) = 9x^2 - 12x + 4$.

(ii) A rectangle x cm wide is such that its length is 5cm longer than its width. Write expressions, in terms of x, for

(a) The length   (b) The area

(c) The perimeter of the rectangle

5. (i) Joan left school at 10.00 a.m for her home 8 km away walking at the rate of 6 km per hour. She spent half an hour at home. She was then given a ride back to school travelling at 16 km per hour.

a)   Determine the time she arrived their house.

b)   Determine the time when she took off from the house to go back to school

c)   Determine the time she got back to school

d)   Draw the travel graph of her journey

e)   Using your graph or otherwise, determine how far she was away from the school at 12:00 noon. Express your answer to 1 decimal place

(ii) Give the straight lines

$$A: y - 3x - 5 = 0$$
$$B: x = 8 - 3y$$
$$C: 2y = 5 - 6x$$

Determine which of the lines (if any)

a) Are perpendicular
b) Are parallel
c) Pass through the origin

6. The table below shows the group frequency distribution of examination marks for 120 students. Each mark is a whole number

Marks	Number of candidates
1-10	0
11-20	2
21-30	6
31-40	7
41-50	14
51-60	20
61-70	35
71-80	29
81-90	6
91-100	1

Construct a cumulative frequency table for this frequency distribution. Take the first class to be ≤ 10.

(a) Candidates with a score of 50 or less will have to resit the examination and those with scores above 60 will be given credit certificates.

Using our table or otherwise, determine, the number of candidates who

(b) will have to resit for the examination.

(c) Earned credit certificates

(ii) Bih has 6 red earrings and 10 black earnings in her box. In the dark, she takes two earrings a random, one after the other without replacement. Find the probability that she took out

(a) A pair of red earrings

(b) A pair of black earrings

(c) Two earrings of different colours.

7. (i) To answer this question, use only a ruler, pencil and a pair of compasses.

    (a) Draw a line segment, $PQ = 7$ cm long in the middle of a new page.

    (b) Bisect line PQ and label the mid-point, X

    (c) Draw a circle such that PQ is the diameter

    (d) Locate a point A on the circumference of the circle such that $AP = 4.5$ cm.

    (e) Draw and measure line AQ

    (f) Write down the value of angle PAQ

(ii) The transformation T is such that $T: \begin{pmatrix} x \\ y \end{pmatrix} \longrightarrow \begin{pmatrix} x' \\ y' \end{pmatrix}$, defined by

$$\begin{pmatrix} 0 & 1 \\ -1 & 0 \end{pmatrix} \begin{pmatrix} x \\ y \end{pmatrix} + \begin{pmatrix} 3 \\ -2 \end{pmatrix} = \begin{pmatrix} x' \\ y' \end{pmatrix},$$

    (a) The point (h, k) is mapped by T onto $(7,4)$. Find the values of h and k.

    (b) Find the point which is unchanged under the transformation

8. (i) The following shows a TV with power button P, connected to a socket S. Using 1 for on and 0 for off, draw the truth table for

    (a) $S \wedge P$ (b) $S \vee P$

    (c) State the relationship that exists between $S \wedge P$ and $S \vee P$.

    (d) Give a counter example to show that this relationship is not always true of any two statements S and P?

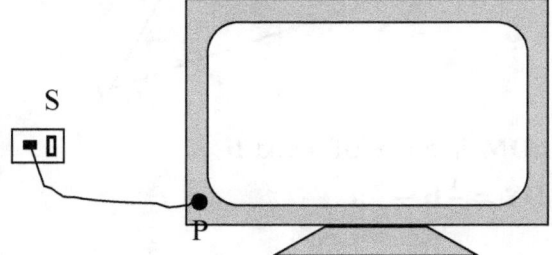

(ii) A ship takes off from harbour H and sails due East. After travelling 15 km it reaches a point P. At P the ship turns and heads due south to reach a point, S, where it then stops. The distance from P to S is 20 km

a) Draw a diagram to show the journey of the ship from H to S. Calculate

b) How far the ship is from the starting point when it stops

c) The distance the ship had travelled before coming to rest.

d) The bearing of H from S, to the nearest degree

9. (i) The position vectors of points A and B are respectively $4i - 6j$ and $-2i + 8j$, where i and j are unit vectors along the x-and y-axes respectively. Given that O is the origin and P and Q are the midpoints of OA and OB respectively

(a) represent this information in a diagram

(b) express vectors **AB** and **PQ** in a similar manner

(c) state the relationship between vectors **AB** and **PQ**

(d) find $\left|\frac{1}{2}\mathbf{OA}\right|$.

(ii) Given that $\mathbf{FD} = \mathbf{a}$ and $\mathbf{FE} = -\mathbf{b}$, the point M is on DE such that $DM = 2ME$ and Y is the mid-point of FM.

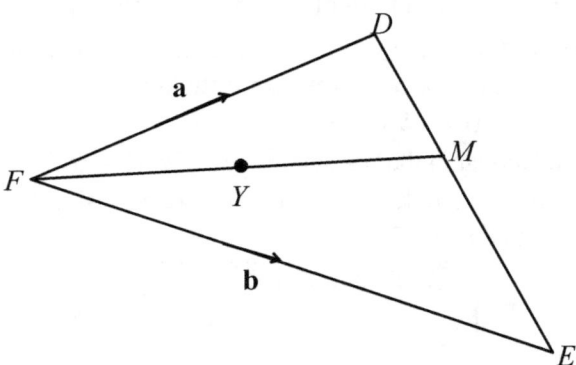

a) find vector **DM** in terms of **a** and **b**

b) show that $\mathbf{DY} = \frac{1}{6}\mathbf{b} - \frac{4}{3}\mathbf{a}$

Answers to Examination Type Questions

ANSWERS TO EXAMINATION TYPE QUESTIONS

MULTIPLE CHOICE QUESTIONS 1

1.	D	2.	C	3.	A	4.	B	5.	B
6.	C	7.	D	8.	A	9.	C	10.	A
11.	B	12.	D	13.	A	14.	C	15.	D
16.	B	17.	B	18.	C	19.	A	20.	D
21.	A	22.	B	23.	D	24.	C	25.	D
26.	C	27.	B	28.	A	29.	B	30.	C
31.	A	32.	D	33.	D	34.	B	35.	C
36.	A	37.	B	38.	A	39.	D	40.	C
41.	C	42.	B	43.	B	44.	D	45.	B
46.	A	47.	B	48.	C	49.	D	50.	A

MULTIPLE CHOICE QUESTIONS 2

1.	B	2.	D	3.	B	4.	A	5.	D
6.	A	7.	A	8.	A	9.	D	10.	C
11.	D	12.	A	13.	D	14.	B	15.	C
16.	A	17.	D	18.	C	19.	D	20.	B
21.	C	22.	B	23.	C	24.	A	25.	D
26.	A	27.	B	28.	C	29.	C	30.	B
31.	D	32.	B	33.	C	34.	B	35.	D
36.	A	37.	A	38.	C	39.	B	40.	D
41.	B	42.	A	43.	D	44.	B	45.	C
46.	C	47.	A	48.	B	49.	B	50.	C

STRUCTURAL QUESTIONS 1

1. $a = 2$, $b = -1$

2.

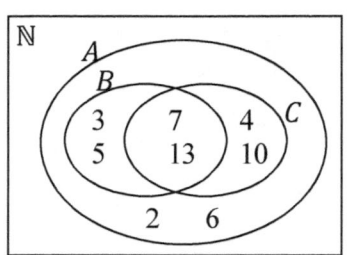

3. (a) $AB = 8$ cm (b) $CD = 30$ cm   4. $50\sqrt{3}$ cm^2

226

5. $\mathbf{OD} = \frac{1}{2}(\mathbf{b} - \mathbf{c})$, $\mathbf{OG} = \frac{1}{3}(\mathbf{a} + \mathbf{b} + \mathbf{c})$ 6. $\angle ABC = 60°, \angle ADB = 108°$

7. 89.8 cm^2 8. $M = \begin{pmatrix} 3 & 0 \\ 7 & 5 \end{pmatrix}$ 9. $y = 24$

10. $3 * 5 = -30$. The operation $*$ is not commutative.   11.   1

12.

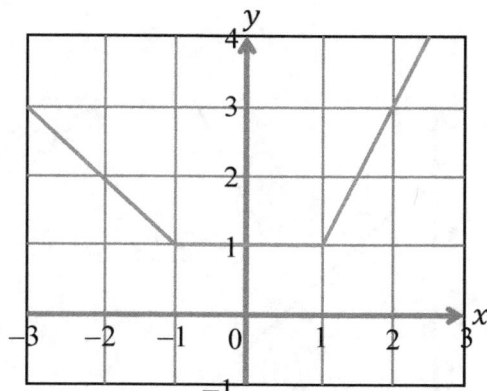

13. $\frac{1}{16}$ 14. (a) $\frac{12}{13}$ (b) $\frac{5}{12}$ 15. 3×10^4 16. $w = \frac{bTu}{a - bT}$

17. (a) 31, 42. Adding the even numbers 2,4,6,8... in succession.

(b) 720, 5040. Dividing the terms by 2, 3, 4, 5,6,7,8 ... in succession.

18. $m = 11$ or $m = -4$ 19. 11_{five} 20. $\frac{r}{s} = 3$ or $\frac{r}{s} = 1$

21. $fg: x \longmapsto 6x^2 + 1$ 22. $\frac{17}{35}$

23. (a) Answers may vary

(b) Line cuts x and y-axes at $(5,0)$ and $(0, -3)$ respectively.   (c) $m = \frac{3}{5}$

24. $< OAC = 45°$.

25. Ratio of the weights of equal volumes of alcohol and water is 19:24.

STRUCTURAL QUESTIONS 2

1. (a) $\{5,6\}$ (b) $\{1,2,3,4,5,6\}$ 2. $r = 14$ cm, $C = 88$ cm

3. (a) $\{x: x = 1\}$ (b) $\{x: x = -1, 1\}$ (c) $\{x: x = -\frac{4}{3}, -1, 1\}$

4. $-\frac{4}{3}, -\frac{1}{2}, \frac{1}{3}, \frac{3}{4}, 1$ 5. 8,3,0, $-1,0$ 6. 56°   7. $\begin{pmatrix} 11 & -7 \\ -16 & 10 \end{pmatrix}$, -2

8. $\mathbf{PM} = \mathbf{r} - \frac{1}{2}\mathbf{p}$, $\mathbf{TM} = \frac{1}{3}\mathbf{r} + \frac{1}{2}\mathbf{p}$ 9. 10 cm   10.   (b) and (h)

11. (a) 5 km^2 (b) 1: 200,000   12. (a) 60 m/s   (b) P starts from rest

because when $t = 0, v = 0$. When p again comes to rest, $t = 64$ seconds.

13. (a) Nigeria is an African county. Ghana is an African county.

(b) Mr. Suh will sign the document. Mr. Suh will be dismissed.

14. (a) $24 - m$ (b) $\frac{m+6}{30}$ 15. $a = 2, b = 0, c = 0, d = 1$ 16. 40.4 m

17. $x = 30°, y = 60°, 90°$ 18. $A = \frac{\pi n}{vV} + 3$ 19. (2,8), $\begin{pmatrix} 1 & 2 \\ 1 & -1 \end{pmatrix}$

20. $-1 < x \le 4$

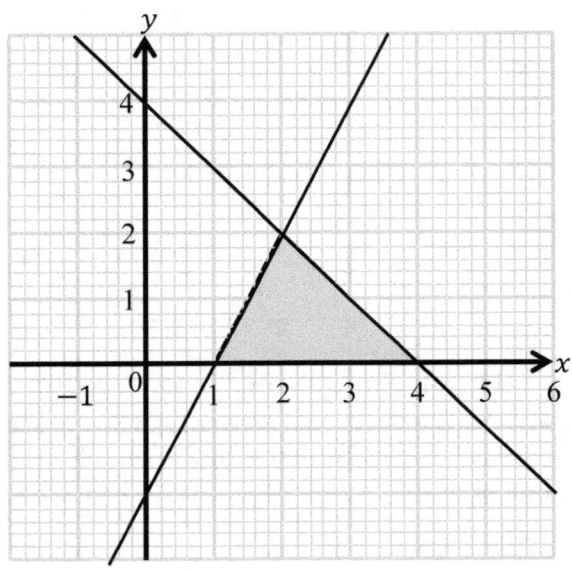

22. -130 23. $-40°$ 24. $\begin{pmatrix} -\frac{1}{8} & \frac{1}{4} \\ \frac{3}{4} & -\frac{1}{2} \end{pmatrix}$ 25. 0.0075

STRUCTURAL QUESTIONS 3

1.   9   2.   600,000 FCFA   3.   300 m/s
4.   7   5. (a) 5,11,22   (b)   356   6. $a = 4, \ b = -1, \ c = 5$
7. (a) $f^{-1}: x \longmapsto \frac{1-3x}{x}$ (b) $ff(x) = \frac{x+3}{3x+10}$
8. 992800 FCFA   9. $-3 < x < 1$
10. $AB: y = 3, \ AC: x = 2, \ BC: x + y = 7, \ x > 2, y > 3, \ x + y < 7$
11. $t = r\left(\frac{p}{a-p}\right)^2$ 12. (a) 1.806   (b) 0.301   (c) 0.155
13. (a) minor arc   (b) radius   (c) isosceles (d) chord (e) minor sector
14. (i) b (ii) c (iii) b (iv) c
15. The value 3 is wrong. The correct value is 2.25   16. $S = 1$

228

17.

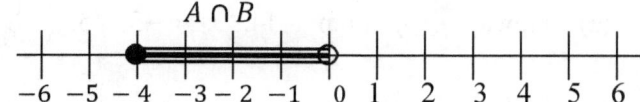

18. (a) 14 cm   (b) 44.1 cm^2 19. $x = 73°$, $y = 92°$

20. $\begin{pmatrix} -2 & -2 \\ 2 & 2 \end{pmatrix}$ 21. $\mathbf{b} = 3\mathbf{a} - 2\mathbf{c}$

22. (a) B and C are parallel   (b) A and D are perpendicular
    (c) None passes through the origin   (d) C passes through $(2,1)$

23. 321 cm^2

24. $A = \{3,5,7,9, ...\}$, $B = \{1,7,17,31, ...\}$, $C = \{-1, 2, -3, 4, ...\}$

25.

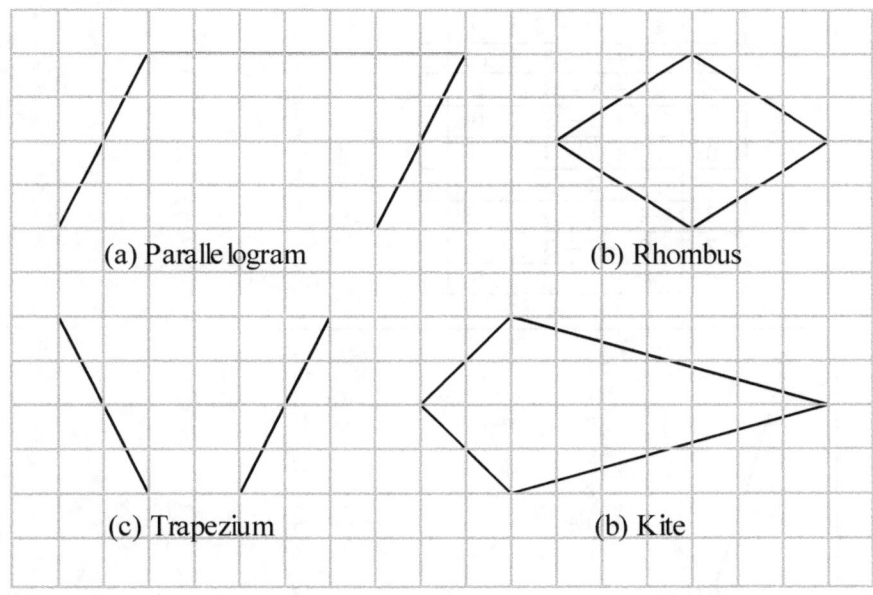

(a) Parallelogram   (b) Rhombus

(c) Trapezium   (b) Kite

ESSAY QUESTIONS 1 (NWR-Mock 2015)

SECTION A

1. -7 2. (i) John is not hungry   (ii) John is hungry and thirsty
    (iii) $\sim p \wedge \sim q$ or $\sim(p \vee q)$ (iv) $p \Rightarrow q$ 3.   0

4. $\dfrac{11}{5}$ 5. (a) 1   (b) 2   6. -4 or 8 7. $x = 20°$, $y = 30°$

8. $x < -5$ and $x > 1$

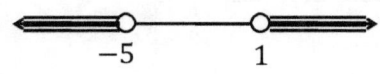

9. $A = \{(1,2), (3,1), (3,2), (4,3)\}$ 10. $\dfrac{7}{31}$ 11. (a) $\left(\frac{1}{2}, -2\right)$ (b) $\sqrt{113}$

229

12. $x = 3$, $y = 4$ 13. (a)   Ngwa   (b)   Lum   14. $x = -3$ 15. $\begin{pmatrix} 8 \\ -9 \end{pmatrix}$

SECTION B

1. (i) (a) Discount on cooker with life span 54 months = 22275 Frs.
        Discount on cooker with life span 72 months = 29700 Frs.
        Discount on cooker with life span 72 months = 32175 Frs.
    (b) Displayed price = 202,5000 Frs
    (c) Amount paid = 180,225 Frs.
  (ii) (a) $P(x) = 3(x + 1)(x + 2)$ (b) $P(x) = 3x^2 + 9x + 6$
    (c) $x = -1$ or $x = -2$
2. (i) (a)   45   (b) $gf(x) = x^2 - 2x - 3$

(c)

x	$x^2 - 2x - 3$	y
-3	$9 + 6 - 3$	12
-2	$4 + 4 - 3$	5
-1	$1 + 2 - 3$	0
0	$0 + 0 - 3$	-3
1	$1 - 2 - 3$	-4
2	$4 - 4 - 3$	-3
3	$9 - 6 - 3$	0
4	$16 - 8 - 3$	5

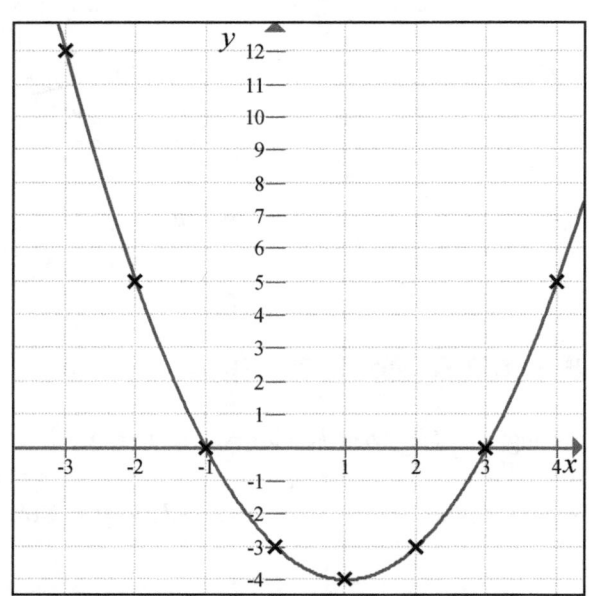

From the graph -1 or $x = 3$
  (ii) (a)   480 tiles   (b)   192,000 Frs.
3. (i) (a) f (b) b (c) c (d) g (e) $c \cup g$ or c and g

(ii) (a) $N = k\sqrt{t}$ (b) $k = \frac{1}{3}$ (c) $t = 225$

4. (i) (a)

Marks, x	f
1	5
21	15
41	9
61	6
81	5

(b)

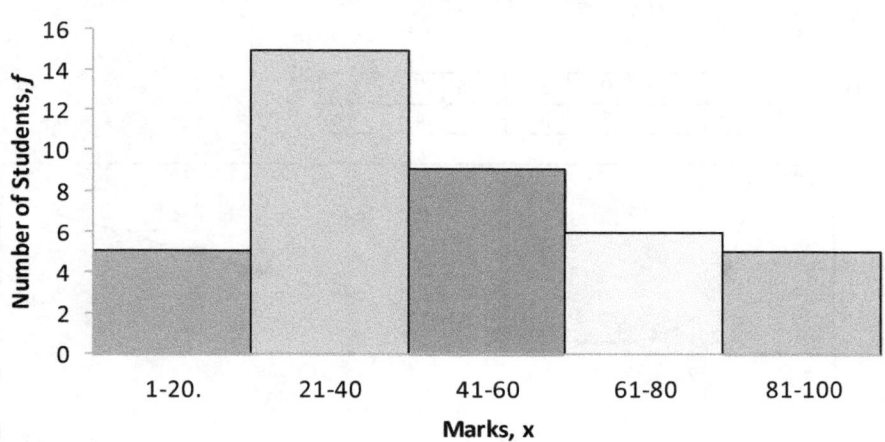

Marks Scored by 40 Students

(c) 46

(ii) (a) $\frac{9}{25}$ (b) $\frac{13}{25}$

ESSAY QUESTIONS 2 (CGCE - 2004)

1. (i) (a) $f(-2) = 1$ (b) $f \circ g: x \longmapsto x^2 + 4x + 1$
    (c) $g^{-1}: x \longmapsto x - 2$ (d) $p = 3$ and $q = -1$
    (i) $x = 7, y = 8$

2. (i)   (a) $\sim p$ (b) $p \wedge q$ (c) $\sim p \wedge \sim q$ or $\sim (p \vee q)$
    (d) $p \vee q$ (e) $p \wedge \sim q$ (f) $\sim (p \wedge q)$
  (ii)   (a) No student does both Biology and History.
    (b) All students who do Biology do Geography.
    (c) $G \subset \mathscr{E}$ (d) $n(B \cap G \cap H) = 0$ or $B \cap G \cap H = \emptyset$
    (e) $n(H \cap G) = 40$.

3. (i) (a) $\mathbf{AB} = \mathbf{b} - \mathbf{a}$ (b) $\mathbf{AM} = \frac{1}{4}(\mathbf{b} - \mathbf{a})$ (c) $\mathbf{OM} = \frac{3}{4}\mathbf{a} + \frac{1}{4}\mathbf{b}$

  (ii) (a) $2\mathbf{p} = \begin{pmatrix} 6 \\ 4 \end{pmatrix}$ (b) $|\mathbf{q}| = \sqrt{41}$ units   (c) $2\mathbf{p} - \mathbf{r} = 8\mathbf{i} + 5\mathbf{j}$

(c) $t = 3,\ k = 2$

4. (i) (a) 21600   (b) 26.7%   (c) 350   (d) 3,150,000 FCFA   (ii) 200 m

5. (i)   (a)   (b)   68   (c)   68   (d)   68   (e)   68.3

x	f
66	2
67	3
68	7
69	4
70	3
71	1

(ii) (a) $\dfrac{3}{8}$ (b) $\dfrac{2}{3}$ (c) $\dfrac{1}{4}$

6. (a)

x	0.25	0.5	1.0	1.5	2	2.5
f(x)	4.5	3	3	3.7	4.5	5.4

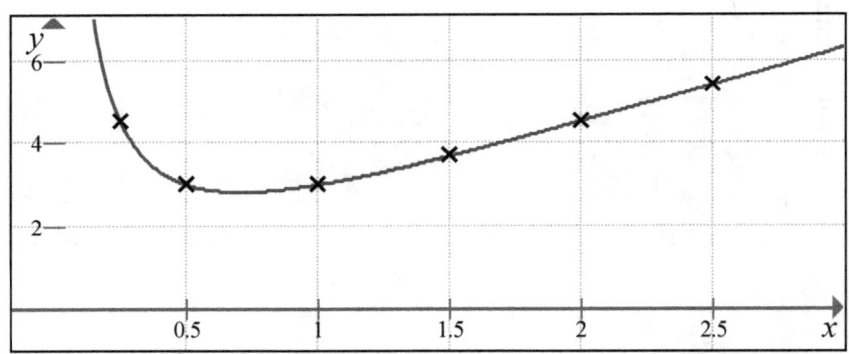

(c)   2.6   (d) $m \approx 4$ (e) $x \approx 0.4$

7. (a) the coordinates are $A'(1,-1),\ B'(3,-2),\ C'(4,4)$

(c) Rotation through $-90°$.   (d) $T^{-1} = \begin{pmatrix} 0 & -1 \\ 1 & 0 \end{pmatrix}$, Rotation through 90°.

8. (i) (a)

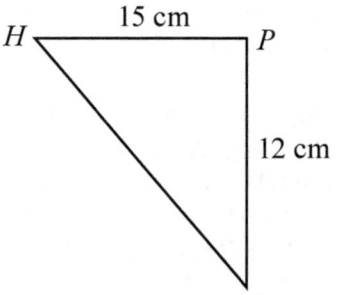

(b)   29.2 km
(c)   70°

(ii) (a) 0°   (b) 75°

(iii) $x = 78°$
        $y = 53°$
        $z = 24°$

9. (i) (b) 8 cm   (c) 1:16
   (ii) (a) $f(1) = 0$ (b) $f(x) = (x-1)(x+3)(x+4)$

232

ESSAY QUESTIONS 3 (CGCE - 2005)

1. (i) (a) 60 l (b)   12500 FCFA   (c) 5300 FCFA
   (ii) (a)   42000 FCFA   (b)   33600 FCFA

2. (a)   12   (b)   12   (c)   15   (d)   24   (e)   19

(f)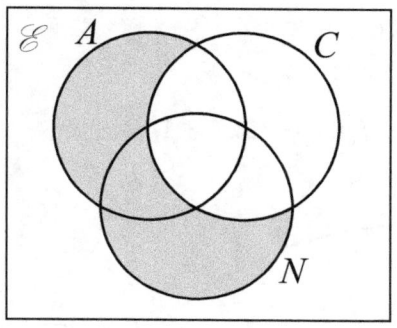

(g)   Those who tuned to both Abakwa FM and Afrique Novelle but not to CRTV.

3. (a)

x	-2	-1	0	1	2	3	4	5
$f(x)$	-8	-2	2	4	4	2	-2	-8

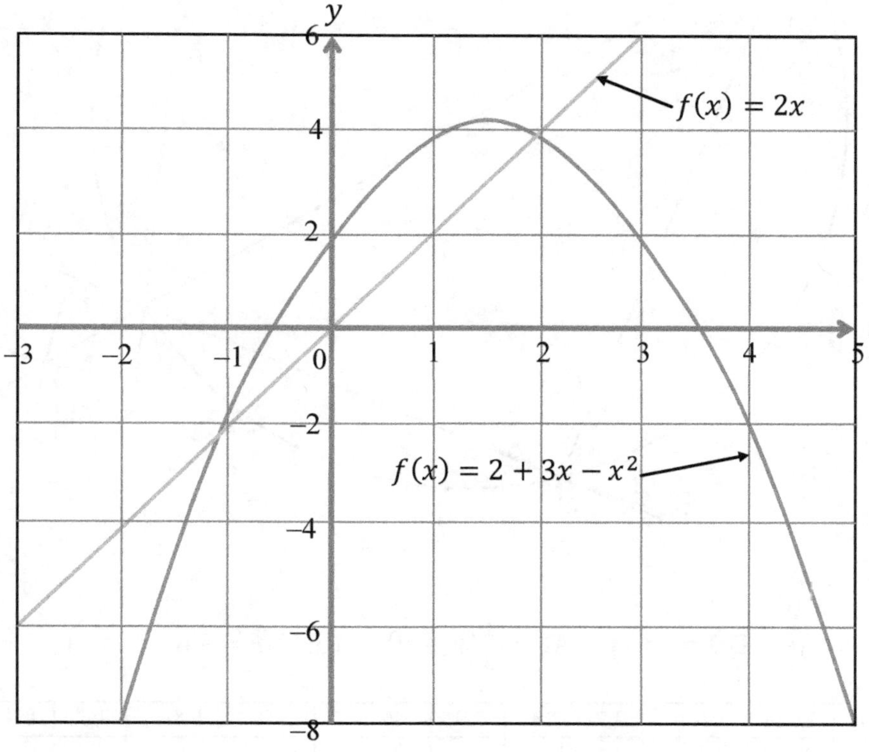

$f(x) = 2x$

$f(x) = 2 + 3x - x^2$

(b) Maximum value of $f(x) = 4.25$ corresesponding to $x = 1.5$.

(c) From the graph $x = 0$ or $x = 2$.

4. (i) $x = 81°$, $y = 138°$, $z = 88°$

  (ii) (a) $p = 9$, $q = 7$ (b) The other factors are $(2x - 1)$ and $(x + 3)$

5. (i) (a) 5   (b) 8   (c) 5   (d) $FE = 4$ cm, Area $= 48$ cm^2

  (ii) $N = \begin{pmatrix} -2 & 3 \\ 3 & -4 \end{pmatrix}$, $a = 7$, $b = 8$.

6. (a) $f(-3) = -7$, $h(-2) = \frac{1}{6}$

  (b) $h^{-1}:x \mapsto \frac{4x-1}{x}$, $x \in \mathbb{R} - \{0\}$, $fg:x \mapsto 2x^2 - 1$, $x \in \mathbb{R}$

  (c) $x = 1$ or $x = 0$

7. (i) (a) – (e)

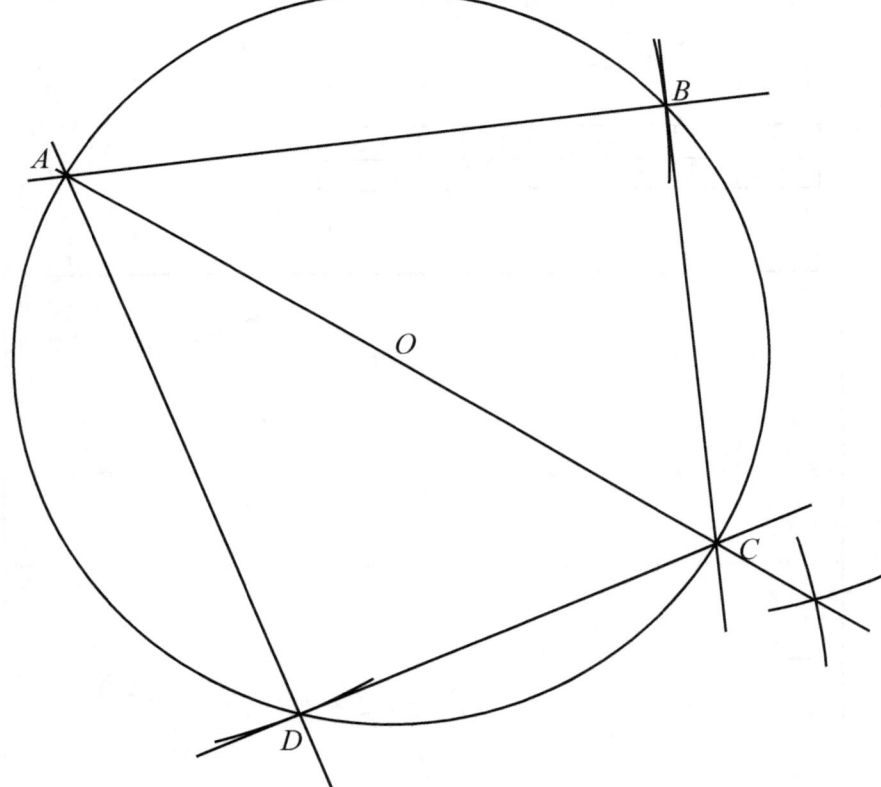

  (f) 8 cm.

  (ii) (a) $DB = u - v$, $AF = \frac{1}{3}(u + v)$ (b) $BF = \frac{2}{3}BE$

8. (i)

Length, x (cm)	7.7	7.8	7.9	8.0	8.1	8.2	8.3	8.4
Tally	‖	‖‖	⦀⦀	⦀⦀ ⦀⦀ ⦀	⦀⦀ ⦀⦀ ‖	⦀⦀	‖‖	⦀
Frequency	2	4	6	13	12	6	4	3

(a) 8.0 cm   (b) 8.01 cm   (c) 8.06 cm   (ii) $\dfrac{2}{3}$

9. (i) 283.56 m   (ii) (a) 288.70 m   (b) 134.61 m   (c) 270°

ESSAY QUESTIONS 4

1. (i) (a) 44000 FCFA   (b) 225000 FCFA   (c) 5000 FCFA   (d) 2.3 %
   (ii) (a) $x^2 + 3x - 18 = 0$ (b) $x = --6$ or $x = 3$
2. (i) $x = 3$, $y = 29$, $z = 14$

   (ii) (a) $\mathbf{B}^T = \begin{pmatrix} 5 & -2 \\ 6 & 3 \end{pmatrix}$ (b) $A^1 = \begin{pmatrix} 1 & -\frac{1}{2} \\ 1 & -\frac{3}{4} \end{pmatrix}$ (c) $3\mathbf{A} - \mathbf{B} = \begin{pmatrix} 4 & -12 \\ 14 & -15 \end{pmatrix}$

   (iii) $\mathbf{M} = \text{from}\ \begin{array}{c} \\ P \\ Q \\ R \end{array} \begin{array}{c} \text{To} \\ \begin{array}{ccc} P & Q & R \end{array} \\ \begin{pmatrix} 0 & 2 & 1 \\ 1 & 0 & 0 \\ 1 & 1 & 2 \end{pmatrix} \end{array}$

3. (i) (a)

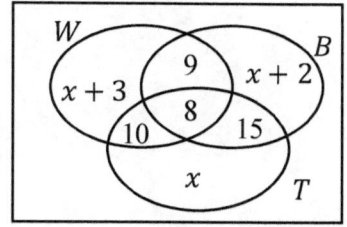

   (b) $x = 11$ (c)   14   (d)   15
   (ii) (a) $A \cap B'$ (b) $R \cap Q \cap P'$ (c) $A \cap B = A$
4. (i) (a) The domain of g excludes because if $x = 4, g(x)$ is undefined.
   (b) $-\dfrac{1}{4}$ (c) $g^{-1} : x \longmapsto \dfrac{1}{x} + 4$ (d) $k = 1$, $m = -6$
   (ii) (a) $l = (x + 5)$ cm   (b) $A = x^2 + 5x$ (c) $p = 4x + 10$
5. (i) (a) 11:20 a.m.   (b)   11:50 a.m.   (c)   12:20 p.m.

(d)

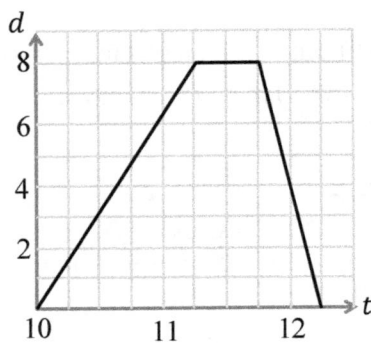

(e) 4 km away from the school.

(ii) (a) $A \perp B$ (b) $A \parallel C$ (c) none.

6. (i) (a)

Marks	f	f_c
≤ 10	0	0
≤ 20	2	2
≤ 30	6	8
≤ 40	7	15
≤ 50	14	29
≤ 60	20	49
≤ 70	35	84
≤ 80	29	113
≤ 90	6	119
≤ 100	1	120

(b) 29   (c) 71

(ii) (a) $\dfrac{1}{8}$ (b) $\dfrac{3}{8}$ (c) $\dfrac{1}{2}$

7. (ii) (a) $k = 4, \quad h = -2$ (b) $\left(\dfrac{5}{2}, -\dfrac{1}{2}\right)$

8. (i) (a)

S	P	$S \wedge P$
1	1	1
1	0	0
0	1	0
0	0	0

(b)

S	P	$S \vee P$
1	1	1
1	0	0
0	1	0
0	0	0

(c) $S \wedge P = S \vee P$ (d) Answers may vary.

(ii) (a)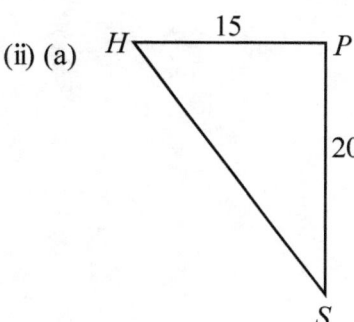

(b)   25 km   (c)   35 km

(d)   323°

9. (i) (a)   See graph on next page
   (b) **AB** $= -6\mathbf{i} + 14\mathbf{j}$,   **PQ** $= -3\mathbf{i} + 7\mathbf{j}$ (c) **AB** $= 2\mathbf{PQ}$
   (d) $\sqrt{13}$ units
   (ii) (a)   **DM** $= \frac{2}{3}\mathbf{b} - \frac{2}{3}\mathbf{a}$

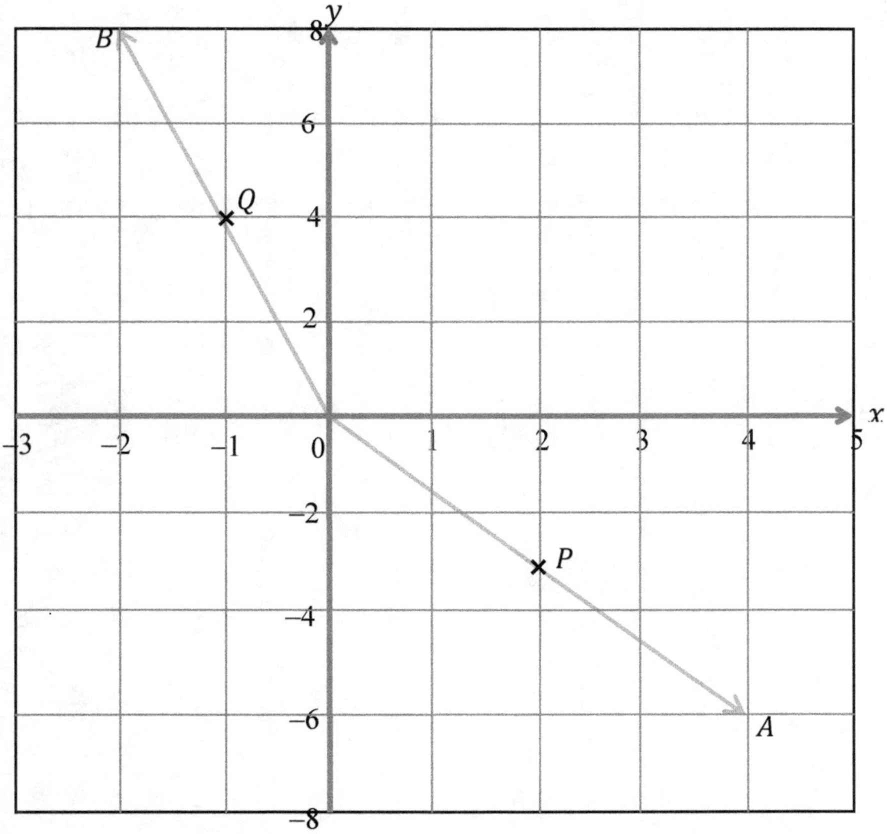